OTHER
SENSES,
OTHER
WORLDS

DORIS JONAS *and* DAVID JONAS

OTHER SENSES, OTHER WORLDS

 STEIN AND DAY/*Publishers*/New York

First published in 1976
Copyright © 1976 by Doris Jonas and David Jonas
All rights reserved
Designed by David Miller
Printed in the United States of America
Stein and Day/Publishers/Scarborough House,
Briarcliff Manor, N.Y. 10510

Library of Congress Cataloging in Publication Data

Jonas, Doris.
Other senses, other worlds.

1. Life on other planets. 2. Senses and sensation. I. Jonas, Adolphe David, 1913–   joint
author. II. Title.
QB54.J58   152.1   75-11816
ISBN 0-8128-1841-5

CONTENTS

47163

ONE

Where Speculation Begins

ONE OF THESE DAYS OUR descendants, near or distant, are going to find life in some form on other planets either in the solar system, in other parts of our galaxy, or in other galaxies.

The very fact that life has arisen on our earth is evidence enough that it must exist in other parts of the universe, for the elements of which the entire universe is composed are remarkably uniform. If some of these elements have combined in ways that produce life here in our solar system, they must, by the laws of chance and probability, have combined in analogous ways elsewhere. Even in our own galaxy there must be thousands of other planets sustaining life in some form, and all the forces of reason would suggest that it cannot be otherwise in other galaxies.

What miracles of chance and combinations of chances made it possible for life to evolve here on earth?

For life to arise on any planet, certain factors have to be present in certain combinations. The solar system of which the planet is a part must have formed in a way that some or one of its evolving planets takes shape at a suitable distance from the blazing inferno of its central sun—neither so near that its surface temperature inhibits life by intense heat, nor so far that life cannot arise because of insufficient solar radiation. The masses into which the swirling gases originally solidify must be within the range that permits a force of gravity sufficient to hold and retain an atmosphere, since without a protective atmosphere solar radiation would be too intense for life forms to be sustained even if all other elements of life were present.

The chance that these two factors alone—distance from the center and the degree of mass that governs gravity—occur together in just the right circumstances puts a preliminary limitation on the possibility for life. Even after this has occurred, there must be a further series of chances following and working upon chances—and these again interlocking with other chance happenings—so that atmosphere, water, rocks, and soil come into being in states that can form a basis for the evolution of life. Of course, there is also a possibility, remote as it seems, that some form of life might arise on a lightless planet should that planet be capable of generating heat of its own within a life-sustaining range.

The information we have been able to gather from our own solar system suggests that ours is the only planet around our sun that sustains life. It could be that other suns, even thousands or tens of thousands of other suns, sustain no life at all on the planets that circle them. Given the myriad suns in our galaxy and the multiple myriads in the universe, it is impossible to believe that chances similar to those that occurred on earth have not also occurred on many other planets.

Once these miraculous chances have come about, however, the prerequisites for life are rather minimal: an element capable of forming self-replicating chains, like carbon, and another capable of combustion, like oxygen. These, together with hydrogen and nitrogen, form a matrix that may merge with other elements to create all the varied, complex, and wonderful forms of life on earth—from amoebas and bacteria to plants and spiders and fishes and man.

What shapes may life have taken in other worlds? Have they developed into intelligent creatures and, if they have, what sort of intelligences have evolved? Have other kinds of life developed high orders of intelligence capable of developing technologies, and, if so, what sort of technology has arisen from their special kinds of being? Shall we ever be able to communicate with these beings, if they exist, in any meaningful way?

One thing is certain: we have no reason to assume that evolutionary forces on other planets will produce forms or intelligences that are the same as ours, even though the basic raw materials must be similar. Whatever chance factors combine to produce any form of life, infinitely more must combine to produce an advanced form.

Genetic inheritance is only a beginning. Two offspring of the same parents by chance born in different environments will produce eventual descendants so markedly different that after many generations it will hardly be possible to realize that the ancestors of each line had parents in common.

The variables of habitat, the chance availability of mates, natural selection, and sexual selection among the offspring will all have combined and recombined to produce members of the same species as varied as a pygmy, a Watusi, a Swede, a Chinese, or a Melanesian, and eventually to divide into species, much as man, apes, and monkeys have separately descended from the same stem.

Our own earth provides an illustration of the almost incredible number of living forms that can possibly be derived from single-celled organisms that were once the triumph of evolution on our planet. All the species now extinct and all those still flourishing form only a part of the total possibilities, for who knows how many new species will yet take shape?

Suppose our distant ancestors had not had to defend themselves against some wild animals and pursue others for food, and had faced circumstances where a sense of smell was more important than sight or hearing. In that case we should by now have had a vastly heightened olfactory sense. All our intelligence, knowledge, and awareness of the world around us would be based chiefly on that perception. Our sight, by now, would be a trifle less keen.

How can we gain some remote idea of what kinds of being we may someday enounter in the course of interplanetary and intergalactic exploration? Surely the most logical way is to try to imagine what might be the result of some other chance circumstances that could produce different combinations of those senses and intelligences that exist right here on our own earth.

We seldom stop to think about how many worlds exist within our own. Influenced by our subjective experience of it, we believe we know it, but to each of its myriad species our world is a different place. Each perceives it differently, sees things we do not see, and is ignorant of things that are apparent to us. If they could describe it to us, they would present such a different picture of it that we should find it hard to recognize it as our own planet. Were we to perceive it through their senses, it would appear to exceed the wildest inventions of science fiction.

For instance, while our own vision is limited by the surfaces of the objects we see, sound penetrates surfaces to different depths, depending upon the nature of the material of which the objects are composed. To a creature that perceives its surroundings by acoustical senses rather than by sight, the everyday world we see about us would be a totally strange place, as its world would certainly seem to us.

Indeed, we have all the elements of science fiction right under our noses, and even within us. When we realize all the potentialities of some of our sensory capacities as well as those of earth's other animals, we hardly have to invent anything different for outer space.

Every living species has its own way of receiving information from its environment and of using that information for its own purposes—in other words, its own way of knowing and its own kind of intelligence. If we examined the nature of the perceptions of others of earth's creatures and tried to imagine our knowledge expanded or altered by only some of those other kinds of perception, what would our view of the world be? What would be the nature of our intelligence if it were based on this other kind of knowledge? What other powers, other technologies, would be open to us?

How, ultimately, do we know what we know? Our awareness of the external world and all the conclusions we draw from this awareness, our knowledge, the cumulative storehouse of the product of the awarenesses and intelligences of all individual existences since our species took shape, and the technology based on this knowledge, are founded on the evidence of our senses.

But our senses are limited instruments. Compared with some of those of other creatures, they are crude. The colors we see are confined to that portion of the spectrum, red to violet, that our eyes and brains are able to perceive. Using instruments to extend our natural perceptions, we have become aware that the spectrum goes beyond these limits at both ends.

There are other creatures that perceive different portions of it—that can see the colors beyond violet but not so far as orange or red. How would the world look to us if our capacity for vision had these other limits?

Similar extensions can be applied to any of our other sense receptors. Suppose we could hear as well as a dog, perceive odors as well as a moth. Not only would the world look different to us; it would also sound and smell different, and we should be aware of a different (and in some ways, perhaps, of a far greater) part of it than we now are. Our sense of what is "real" would be different, as would our body of knowledge and the practical devices we should be able to produce.

We do not even have to resort to imagination to dream of other senses it might be possible for us to possess. The animal world of our own earth is full of types of awareness, and therefore of knowledge, that are beyond our ken.

Creatures living on earth have evolved so that they know only what they need to know to enable them to live and to perpetuate their kinds. Some, especially the higher forms, do have a potential for recognizing and adapting themselves to new information if this becomes necessary in a changing environment. City birds, for example, have learned to recognize milk bottles left delivered on doorsteps as a source of food, and have discovered how to use this information by pecking through the bottle caps. Usually, however, in a stable habitat, new kinds of vital information do not too often present themselves, and a creature prospers by sharpening the senses it possesses rather than by developing new ones.

In this respect, man is not very different from the rest of creation. Our evolutionary course fashioned us, too, to know only what we needed to know. Our five senses bring us as much information as we need of the planet we inhabit and the universe of which it is a part. Every species possesses special equipment, either of anatomy or behavior, that has evolved by selection in tandem with its environment, and that enables it to survive. The giraffe's neck, the fly's eye, or the bat's sensitivity to echoed sound come into this category. But—and this is a big "but"—mankind's special equipment for survival is our "new" brain, which evolved together with our youthful form, and our equally "young" behavior.

Curiosity, seeking, and experimenting are behavioral characteristics of most young mammals. With maturity, however, these attributes disappear and the animals fall into the conventional behavior patterns of their kind. Human beings, on the other hand, remain "young" in this sense all their lives. We are constantly curious; we have a *wish* to know more than we need to know merely for survival, an urge to satisfy our curiosity, to know for the sake of knowing.

In ordinary life, of course, we still perceive only what we need to perceive. Our senses scan their respective scenes and select the aspects of them that concern us; normally only these are registered by our brains. But in our "play"—our arts and sciences—we have discovered ways to extend our senses as well as our limbs by using all kinds of equipment sensitive to parts of reality that without their aid would escape our awareness. The microscope reveals to us the subvisual world; the Geiger counter translates unsensed phenomena into terms that we can perceive.

Clearly, "reality" is not simply that world bounded by the awareness we have of it through our five senses and by the body of knowledge we have built on this basis. During all periods of history, theologians, philosophers, scientists, poets, and artists have brought the powers of their minds to bear on the same question: What *is* the ultimate reality? Their answers have been based on the evidence of their senses, and then modified by their personalities, predilections, preconceived ideas of what is possible or impossible, right or wrong, desirable or undesirable.

Take a most elementary example. We all think we know what a woman looks like; but how different she looks via the eyes and feelings and paintbrushes of a Michelangelo or a Toulouse-Lautrec, a Renoir or a van Gogh, a Burne-Jones or a Soutine, a Gainsborough or a Picasso—or, for that matter, via the thoughts and feelings and pens of a Wordsworth or an Albee, a troubadour or a Wylie.

Given these vast differences of awareness between creatures of the same species and even the same culture, all possessed of the same organs of perception, can we begin to imagine what the world would look like if the range of our knowledge of it were to be in some ways altered or extended?

We have extended our senses with instruments. We know about, say, X rays, even if we cannot perceive them. But how far is it possible for our minds, having evolved as a product of *our* perceptions, to encompass concepts, thoughts, or feelings based on totally *different* perceptions?

In one of our sciences, physics, ideas have been developed in recent years that take bold steps in this direction. Increasingly sophisticated experimentation presented facts that could not be fitted into the reality evidenced by our senses. How could we conceive of light as material and nonmaterial at the same time—of being at once a particle and a wave; of an electron being in two places at the same time—of space being curved? We have had to express these concepts in the abstract logic of mathematical terms, a language totally removed from the evidence of our senses, and one that conveys no image of reality to our minds.

If, indeed, we can shed our present concepts of the nature of reality, perhaps, by chance, we shall hit upon something like the kinds of intelligences that may have arisen in other worlds, consisting of the same elements as ours but perhaps in different proportions, where chance has taken different turnings. Like Robert Frost on his path in the forest, we may speculate on what our course might have been if we had taken the other arm of the fork in the road we traveled, or we may go further and try to reconstruct in our minds the consequences of having taken a totally different road.

Even then our insight must be modified by the stage that those other worlds, and we in the meantime, have reached. Had an enterprising intergalactic traveler reached our planet a hundred million years or so ago—as perhaps he or she did—could that traveler possibly have appreciated the potential it possessed to produce us, its currrent heirs, and our descendants in whatever form they may take?

Each sense we possess gives the equivalent of a narrow slit window on reality. The small part of the material universe we thus experience provides us with data that our brains, our intelligence, must process. By seeing similarities and differences with other perceived data, by making comparisons with data stored in our memories, by drawing conclusions and by checking those conclusions against the stored memories of our race that have been passed on to us by word of mouth and in writing, we arrive at ideas that we believe correspond with reality.

If the slit windows of other creatures elsewhere in the universe are a little wider than ours in some places, a little narrower in others, or if they open onto a different part of the cosmic scene, how different would the conclusions they arrive at be from ours?

Nor is this the only question. We have to look at the nature of intelligence itself.

When we use the word *intelligence*, we want to free ourselves from its immediate connotation of the particularly brilliant human mind. We find that the more we analyze intelligence into its ingredients—assembling, retaining or discarding, and using information—the more we see it as a universal principle in all of life.

Our tendency has been to associate intelligence with the nervous system, and especially with that part of the nervous system that is the most specialized for receiving, interpreting, and using information—the brain. But as we continue our exploration of its nature we find intelligence manifest in many forms, not least in the function of every cell.

For this reason, it is inevitable that we shall find intelligence elsewhere in the universe. The uncertainty is only as to the stage at which we shall find it. Since intelligence evolves along with life, the kind and degree of intelligence we meet will depend upon the stage in its evolutionary course at which we meet it.

Can we find clues that will give us an indication of its course? Is it possible for us to imagine the nature of intelligences higher, and far higher, than ours?

Other Worlds:
The Sense of Smell

WHEN THE UNITED STATES government decided to get man to the moon within a decade, the scientists and technicians assigned to the task had to begin with the information they already had.

At that time the moon was as remote in most people's minds as anything in the infinities of space. All kinds of serious and fanciful imaginings about what might be found there were expressed: childlike fantasies of the "man-in-the-moon," "bug-eyed monsters," and "little green men," scientific fears of bacterial life against which we might have no immunity, quicksand-like surfaces, or other dangers.

The unknown has a way of sparking our imaginations in much the same way as children, afraid of the dark, a closed room, or an unfamiliar forest, people such places in their dreams with monsters, demons, ogres, or fantastic genies or sprites. When the dark place is entered, the room opened, and the forest explored, all these fears evaporate and we deal with what is actually there.

Much the same sort of problem faced the first astronauts and scientists when they began their work in the early 1960s. They began by assembling all the information already at their disposal, and extrapolating from this what they could reasonably expect to find on the moon. Where they could, they tested out their ideas in advance; where they could not, they studied the implications and probabilities of all the alternatives.

As a result of all this groundwork, by the time they made their journey to the moon they were sufficiently well equipped to do so. Although they did find some surprises, of course, as in any exploration into new territories, on the whole the reality was in line with what they had come to expect by studying all the possibilities based on logical, reasonable thinking, and by extensions of knowledge already available.

Probably the greatest surprises are to be found here on earth, if we only discover how to look for them. Not only in other animals but also in ourselves are many marvels of which we are unaware, either because they are not generally known or because they are still to be discovered. Perhaps we shall be able to understand what we may discover in other worlds by referring back to what is present in our own. Even the structure of other intelligences may be very much like ours, whether or not they are grounded in other sensory information.

With the experience of the astronauts behind us, we are inclined to take it for granted that if we one day decide to make such journeys, the technical difficulties of "getting there" will be mastered. The difficulty seems not to be in the getting there, but in imagining and therefore in preparing for what we shall find when we arrive. The greatest obstacle here is our usual way of thinking.

With the best will in the world we find it almost impossible to separate our ideas about other possible forms of intelligent life from our ideas about our own. For all of us there is a great mental effort in grasping a novel concept, far removed from what is familiar to us. We are not able to latch it on to what we know, and so we cannot make associations to connect it with our experience. The radically new idea appears weird and improbable, and we cannot accept it until the smaller steps of the in-between stages have been filled in one by one.

If we are to make the imaginative leap out of the confines of what seems possible and familiar, we have to give our imaginations some bending and stretching exercises. We cannot afford to be so conceited, or anthropomorphic, or plainly simpleminded, as to think that the possibilities of intelligent life are exhausted by our own form of it.

On the other hand, limitations within which we may let our ideas run do exist and are real. All life is tied up with the nature of carbon compounds. It is true that silica can also form long-chains, but these are not so stable as carbon chains and would be an unlikely basis for complex life. As our very first step, then, we can eliminate other elements, and confine our thoughts to what is possible for carbon chains. Since we ourselves, and all the animal and vegetable life we see around us, are examples of the ultimate possibilities of carbon compounds, we have some very solid evidence to guide our thinking.

Let us imagine that somewhere in the infinity of space a planet exists where the conditions of life have led to the highest refinement of the sense of smell. A sense of smell dominates the perceptions of the most highly evolved creatures in that world. Their brains are able to utilize the information brought to them by this means—that is, to construct ideas of their environment, develop a language of symbols, and eventually produce abstract ideas and a technology based primarily on a sense of smell.

We shall call these creatures the Olfaxes, and we are going to assume that they possess the same other senses as we do, but in a more primitive or blunted degree.

On their planet the atmosphere is perpetually murky, so that no great advantage would be gained by creatures possessing particularly keen sight. Selection obviously has preserved and perpetuated those with remarkable acuity of some other senses, in this case, the perception of odors.

On our own planet most higher species receive their information in this way, too, and base their knowledge upon what their brains and nervous reflexes can make of perceived odors.

In the same way that we *see* trees, rivers, mountains, animals, buildings, signs, rocks, mud, sand, and faces, and form an impression of the world around us primarily through these sights, so these species perceive these things primarily by their odors. The other senses act to test, prove, confirm, deny, or modify the impressions of the olfactory one.

Odor-producing chemicals function as the printer's ink of

these species. As we make abstract lines or curves in ink or paint on paper or canvas so that others may see these symbols and interpret them, so most other species of higher animals on earth transmit information to others of their kind by leaving marks of urine, feces, or the products of specialized glands in certain places.

Of all earth's creatures that give and receive social news of each other via the sense of smell, one of the best examples is the wolf. In 1909 E. T. Seton included this description in his *Life Histories of Northern Animals*:

> Incredible as it may seem at first sight, there is abundant proof that the whole of a region inhabited by wolves is laid out in signal stations or intelligence posts. Usually there is one at each mile or less, varying much with the nature of the ground. The marks of these depots, or odor-posts, are various: a stone, a tree, a bush, a buffalo skull, a post, a mound, or any similar object serves, providing only that it is conspicuous on account of its color or position; usually it is more or less isolated, or else prominent by being at the crossing of two trails.

Further on he continues,

> There is no doubt that a newly arrived wolf is quickly aware of the visit that has recently been paid to the signal post—by a personal friend or foe, by a female in search of a mate, a young or old, sick or well, hungry, hunted, or gorged beast. From the trail he learns further the direction whence it came and whither it went. Thus the main items of news essential to his life are obtained by the system of signal posts.

From what we know of the communicative abilities of animals like the wolves, we are made aware of the possibilities of odor as a vehicle for conveying and receiving information. We ourselves subject an odor to analysis and give names to its component elements so that we can describe smell information in terms of chemical formulas. Although these formulas allow us to reproduce the odors in a laboratory, the symbols that comprise them are meaningful only to a chemist. To the rest of us they communicate nothing about the odor—or anything else.

If we endow our Olfax population with a highly developed intelligence and a technology based on this kind of transmission of knowledge, however, they surely will have found methods of codifying scent information. Presumably, the means of transmission they would use would be analogous to our use of language, which is based on our most highly developed sense: sight. We give names to impressions we receive via our senses, assign descriptive words to thoughts, activities, positions in time or space, and so on, and then break up these words into sound elements, each of which we represent by a letter or group of letters. These letters, written on a convenient surface, then provide us with a visible code that we can read and process again in our brain to receive and conceptualize the thought or image encoded in written words.

The Olfaxes will, like us, have invented "words" and "names" for the thoughts and experiences they conceive via their sense of smell. These "words," reduced to a code of odorous particles, can then be suitably recorded so that they can be smelled again when required and these smells reprocessed in their brains to convey the original concept.

Before we describe the way we think they may do this, let us explore some of the many intriguing things about the sense of smell.

Smell signals have a quality that we human beings do not immediately recognize: the large number of elements available for the synthesis of any particular odor. A scientist once tried to devise a machine that would reduce the smell of a human being to its elements, and he found twenty-four separate ones. No doubt a dog would have identified many more; but even so, in all their possible combinations and permutations, just twenty-four separate elements would be enough to identify every single human being in the world with as individual a mark of recognition as his or her fingerprint.

An intelligent race like the Olfaxes, with a sense of smell developed beyond that even of our dogs, and with the kind of brain that could use this information to the fullest extent, would identify a specific odor with every individual they met in the course of their lifetimes and would be able to identify them instantly and far more precisely than we can by name. Names are sometimes duplicated, but a person recognizable by his or her odor carries an unmistakable calling card.

Being biased to perceiving the world visually, we find it difficult to realize that the olfactory sense may be as efficient and pliable as sight, and perhaps capable of even greater modulation. The information we receive through our temperature sense, touch, vibration, balance, and others, is not sufficiently accurate to enable us to use it when exact measurement counts. With all our senses, except, perhaps, only hearing, we can use *with precision* only one—and that is sight and its resultant visual image.

On earth, although much information is passed by scent and many complex and wonderful ways of life are based on such intelligences all of them here are surpassed by information and intelligence, all of them are surpassed by information and hypothetical Olfaxes and how they may have developed, we imagine that because of the environment in which they live, olfactory intelligence has reached the level at least of our own present basically visual intelligence.

To give an idea of how the information received by the Olfaxes and by us might differ, we can take a simple illustration. We can see sun, moon, and stars, and we can draw certain information from these observations. We can then use our brains to supplement this information by logical extensions of it and use these for navigation, for gaining knowledge of the nature of the universe, and eventually for technology based on this knowledge.

The Olfaxes, on the other hand, would not see the sun, moon(s), or stars of their planetary system. Even if they possessed good sight, the murkiness of their atmosphere would preclude this. However, they feel the warmth of the hidden sun, or sense cosmic electromagnetic emanations. Using this information, they will be able to construct instruments that can penetrate their atmosphere and receive signals that they can translate into olfactory terms, just as the instruments of our astronomers can receive electromagnetic and other energy pulses, light waves, and so on, and register these in visual terms.

This idea is by no means as fanciful as it might at first seem. Even among human beings who, by comparison with many other animals, are rather numb to odor signals, we can demonstrate the remnants of the inherent acuity of the olfactory sense.

One of the substances that can be perceived independently by three of our senses is alcohol. We can smell it, taste it, or feel it if it is rubbed on our skin. But it takes sixty thousand times the amount of alcohol for us to feel it on the skin, and twenty thousand times as much to taste it, as it does to smell it. Our taste organs are equipped to perceive relatively large amounts of a substance, but smell brings us news of the presence of the most minute quantities.

And when we remember that this faculty of smell, vestigial and yet still so sensitive in human beings, is magnified a hundred- and a thousandfold in others of earth's creatures, we begin to realize how potent, how malleable, and how refined an instrument for communication it could be if it were at the disposal of highly intelligent beings.

In man the olfactory organ is very small: a mere 5 square centimeters altogether. The airway of the nostril provides access to the sensitive nerve endings all but hidden high in a crypt in the upper reaches of the nasal cavity, and it comprises some 5 million olfactory-sensory cells. The main movement of air in ordinary breathing takes place throughout a high arc in the streamlined nasal cavity; nearly all the current passes through without coming into contact with the olfactory region.

Nonetheless, by way of convection and diffusion, odorous particles escape the main airstream and, like rising smoke, reach the uppermost region where the olfactory organ is located. When we chew food, for example, the motions of the palate and throat create disturbances in the airstream that send odor particles in an upward direction via the rear access of the nasal opening to the back of the throat (the nasopharynx).

When we sniff, we generate swirls of air that transport the odor particles in eddies to the sensitive region with some force.

By comparison with the human's relatively vestigial organ, the olfactory area of a German sheepdog's nose is 150 square centimeters, or thirty times the size of ours, although the animal is considerably smaller. Moreover, it has 220 million olfactory-sensory cells against a human's 5 million.

Even smaller dogs still outrank man in this respect. A dachshund has 125 million cells, and a fox terrier 147 million. However, these figures tell only part of the story, because the ability to detect odors increases exponentially with the additional cells. Experiments using olfactometers indicate that a dog's sense of smell is *one million* times more acute than a human's.

To convey what this actually means in the ability to smell, a writer about animal senses, Vitus Dröscher, gave a telling illustration. He wrote that if the molecules of 1 gram of butyric acid (an element of human and animal sweat) could be made to evaporate evenly in all the rooms of a ten-story office building, then a human would only just be able to register the smell by putting his nose through a door. But a dog would react to the same gram of odorous particles if they were *diluted to fill the air over the entire city of Hamburg up to a height of 300 feet.*.

Many dog owners have conducted their own experiments to confirm the phenomenal tracking ability of their own dogs. George J. Romanes was one such owner. He arranged for twelve men to walk in Indian file so that each of them put his foot into the footprint of the man ahead of him. After a time the party split up into two groups; each group went to a separate hiding place. Romanes's dog was then set free and she was able to track down her master without the slightest hesitation.

The dog's sense of smell is by no means the ultimate in the animal world. The moth's—which we shall shortly be discussing—stands in superiority to the dog's as the dog's does to ours. If such an ability were coupled with a degree of intelligence capable of reasoning, our mind boggles at the amount of material it would have at its disposal.

Interestingly enough, there seems to be no relationship between the chemical formula of an odorous particle and how it smells to us: some compounds of different chemical structure smell alike to us, while we can readily distinguish others of similar chemical structure. But the size and the geometric shape of scent molecules do seem to be among the essential properties that enable us to smell them.

About a decade ago, three American scientists suggested that the ability to smell is afforded by tiny holes of various shapes in the receptor organ, smaller than can be resolved by an electron microscope, and that a correspondingly shaped odor-bearing particle fits into each of these like a key into a lock. They said that we experience spherical molecules as camphor-like; disk-shaped molecules as musky; disk-shaped molecules with a tail as flowery; wedge-shaped molecules as pepperminty; and rod-shaped molecules as ether.

They also suggested that other scents are composed of these odor elements in the way that any shade of color can be mixed from the three primary hues. On the other hand, we have to say that there are exceptions to this and, tantalizingly, every time a theory is proposed to account for the odorous property of substances, there always seem to be some that refuse to fit into the scheme.

Many odorous solutions strongly absorb waves from both the infrared and the ultraviolet portions of the spectrum, and perhaps it is significant that there is a connection between odors and the part of the spectrum that is invisible to us.

When investigating the sense of smell, the creatures most often used as examples are butterflies and moths, because their olfactory sense is as acute as or more acute than that of any other of earth's creatures. A male moth can smell the sex scent signal of a female of his own kind from as far away as 20 kilometers. One male moth, released 1.6 kilometers away from a female, found her in ten minutes—that is, at a speed of about 10 kilometers an hour. Obviously he did not need much time for searching, but flew directly to her.

The stem of a silkworm moth's antenna, only a quarter of a

millimeter thick, contains no fewer than 40,000 hairlike fibers. Of these, 35,000 conduct olfactory signals while only 5,000 pass on other sensory messages to the moth's nervous system. Experiments of elegant delicacy that involved monitoring a single fiber of an olfactory cell with a microelectrode, revealed one surprising similarity between the olfactory and the visual code. It seems that it is possible for three adjoining cells to react to the same scent in quite different ways: by stepping up the sequence of current impulses, by reducing them, or by not reacting at all. Thus a smell ultimately received by the brain is a blending of these three primary responses, analogous to a blending of the three primary colors by the visual cells of our retina.

So the sense of smell ranks on a par with sight, both in sensitivity and in discrimination. In origin it antedates sight, as evidenced by the structure of the receptor cells. Unlike the retinal cells that have separate ganglia (centers from which nerve fibers radiate), each olfactory rod is a combined receptor and ganglion cell. This simpler form, seen in the relatively primitive nervous systems of the lower vertebrates, persists in higher animals only in the olfactory organ. Indeed, it is one of the most primitive cells in the human body.

All scents alert our olfactory receptor cells quantitatively and qualitatively, and the stimulus they evoke is then transmitted directly to the olfactory bulb in the front of the brain without the intervention of any relaying nerve cells. In this, as we have noted, olfaction is unlike any other sense.

There is, moreover, a basic difference between the sense of smell and the sense of sight in that an object perceived by its odor does not have to be present to be smelled, whereas an object out of the range of vision obviously cannot be seen. Odors are conveyed by chemical elements that are given off and remain suspended in the air for varying lengths of time. For example, those of us who live in the country are made powerfully aware of the past presence in our area of a skunk for many days after it has left. We do not have to see it to know that it is, or was, there.

To a creature attuned to interpreting this kind of information, however, much more is apparent than is relayed to us. Such a creature will know more precisely how long ago the skunk was there, in which direction it was traveling, perhaps the sex and state or condition of health of the skunk, and so on. Olfactory communication thus gives information about the immediate past as well as the present; perhaps, with refinement, information about the more distant and very distant past could be gleaned.

Odor signals are also perceptible around corners or in other blind spots, and also in complete darkness, where sight alone would be useless. Because of these considerations, an intelligence based on the olfactory sense must be of a different nature from our own. Where at least the immediate past, and sometimes the more distant past, blends with the present in its immediacy, concepts and basic ways of thinking must be molded by this kind of awareness.

In *our* thoughts and terms of expression, the past is sharply distinguished from the present. A person out of sight is no longer with us. He *was* there, but he *is* so no longer, and out of sight, as the maxim has it, he is probably out of mind. But for the Olfaxes his odor would persist and he would continue to be part of their immediate present long past the time he is in visual range. This would have to influence their patterns of thinking in ways extremely difficult for us to fathom.

How can we imagine a "vocabulary" of smell? Basically all our *visual* symbols are elaborations of the straight line and the curve. From these extremely simple elements, which are all that are registered by our retinas, our brain recognizes combinations that its associative apparatus converts into visual patterns and words to describe them. Color is added by a similar process. When a light ray with a wavelength of 635 to 640 millionths of a millimeter bombards the retina, the brain interprets this and we "see" red. The same light ray bombarding another portion of our anatomy than the retina is perceived according to the sense capacity of that organ. Our skin may perceive it as heat.

A vocabulary of olfactory units would be at least as versatile in a highly developed brain evolved to interpret and use them. Although the olfactory apparatus of the human nose is not sufficiently fine to perceive light or heat directly, it can infer its presence from the odors released by heat in other substances or organisms: hot meat, for example, smells different from cold meat.

The vocabulary of primarily olfactory-intelligent creatures must consist of particles of untold varieties of chemical substances suspended in the air, as we see a multitude of lines and curves of varying colors, lengths, and combinations all around us. The brain then interprets these chemical particles into an olfactory whole that we must compare with what for us is a mental image. If the Olfaxes have vocal powers, they will give a name to that olfactory whole, that particular combination of odors, as the words "palm tree" evoke in our minds a specific image of a certain combination of visual forms.

Imagine ourselves blindfolded, walking in a garden. We smell the odors of violets, roses, lilies, stock, and syringa. These perfumes are associated in our minds with a remembered image and we perceive those odors largely in terms of that visual memory. Even if we are born blind, so that we have no visual memory associated with any of those odors, our brain, determined by its inherited structure, aided perhaps by our sense of touch, will present us with some kind of mental image, even though it will probably be highly inaccurate; we have no words to describe the odor of a violet and we cannot define in any terms of language, but only in formulas, how the chemicals given off by this flower differ in aroma from those exuded by the rose. But a primarily olfactory intelligence could do this. It would have words that would define reality in terms of just such differences.

On our planet many creatures besides those we have mentioned can do this too, if wordlessly. Fishes can differentiate the smells of certain waters and thus be guided to their home rivers at spawning time. As a matter of fact, a fish's sense of smell is so important to it that some smell not only with their noses but also with large areas of their skin. In 1965 Dr. Mary Whitear located olfactory cells in the skin of the gill covers, the ventral area, and the tails of minnows. A few years earlier, in 1961, Dr. S. L. Smith observed a starfish locating clams buried under the sand below the water by means of its fine sense of smell. The olfactory sense of many insects and of warm-blooded animals is no less acute.

In mankind, there is still a notable connection between the sense of smell and our emotions, based on the fact that our emotions are mediated largely by the same part of the brain that governs our sense of smell.

In our prehuman past, the part of our present brain that governs our judgment and reason, the neocortex, was insignificant both in size and function in comparison with those older parts of our brain, the limbic system and the brain stem, that had developed in our earlier mammalian and reptilian prehistory. In that remote ancestral primate past, our forerunners lived in trees where foliage limited the value of acute vision. Like most other mammalians, they relied more on their sense of smell than on sight. The olfactory brain was as well developed then as our visual brain is today.

At one stage in our evolution either the forests receded because of climatic factors, or for some other reason those ancestral creatures left the forest and began to live in open savannah areas. In that new environment, gradually and over long stretches of time, great changes in their form and behavior took place. They began to run and walk in an upright position, their feeding habits changed, and with them their way of life. Most particularly their relative physical vulnerability vis-à-vis the other, more powerful creatures of the savannah promoted a need for a great development of the most promising equipment they had—the new brain, or neocortex.

The new way of life, in which sight rather than smell was at a premium, coincided with a need for reasoning intelligence, and these two faculties, sight and reason, developed in tandem. But for thousands of millions of years before this new development, our species' useful behavior patterns and its most necessary and basic innate responses were formed and refined in association with the sense of smell.

Individual creatures responded to specific odors that initiated mating and breeding, care of the young, and responses to hostility, territory protection and population spacing, food seeking, shelter finding, social interactions, and the rest of the complexes of mammalian behavior. These diverse behaviors were not initiated by a thought, in our sense, like "It's getting dark, I must build a sleeping nest," but by what we would call moods, or emotions, that automatically released the appropriate behavior, in roughly the same way that when we feel hungry we look for food. The lives of our forerunners were regulated by emotions, and these were promoted by the information the brain received—which at that stage was primarily olfactory.

So it is that today our emotional responses are deeply embedded in the older part of our brain that was never eliminated by only overlaid by our newer, reasoning brain. Probably for that reason, our emotions are often powerfully independent of our newer thinking processes, which only sometimes, and usually not very radically, influence our deepest feelings.

If, by some chances other than those that actually occurred, our species had developed its potential for reasoned thinking processes in a different environment, we might have evolved intelligences based on the olfactory faculty along lines that we are now trying to imagine for our Olfaxes. As it is, the importance of aromas in two fundamental elements of our own behavior—in eating and in sex—makes it possible for us to find *some* common ground with creatures motivated by the olfactory sense. The pleasurable anticipation and the whetting of appetite by the odor of food, and the actual physiological processes, like salivation, released by these smells, speak for themselves, as do the arousal and stimulation of sexual interest and responses in many of us by the use of perfumes.

The close connection between sexual function and olfactory perception was recently brought to light in an investigation about the role of the so-called Jacobson's organ, or vomeronasal organ, located within the nasal structures.

It was noted that all four-legged vertebrates possess two separate chemoreceptor organs, indicating that this olfactory differentiation could date right back to the earliest amphibians. Both systems provide pathways to the hypothalamus, a part of the limbic system of the brain, but the vomeronasal pathway is via brain centers identified with the release of sexual behavior. Vomeronasal receptors appear to be more sensitive than nasal receptors to low volatile compounds, and in large mammals there is strong evidence that the vomeronasal organ functions as a specialized receptor for excreted sex hormones, concentrations of which vary according to the female's reproductive status.

Males of nearly all hoofed animals, as well as members of such disparate groups as bats, insectivores, and carnivores, perform a facial grimace called *flehmen* when presented with female urine. This grimace evidently assists the functioning of the vomeronasal organ, because species that can close their external nares without retracting the upper lip do not need to perform flehmen in order to fill the vomeronasal organ. In this manner, through urine analysis by an organ that assesses its specific smell, the approach of estrus in females of their own kind can be detected well before the onset of "heat." When the concentration of the sex hormones in the urine passes a certain level, sexual behavior is released; below the minimum threshold, no response is evoked. Thus the vomeronasal organ may play a major and hitherto underestimated role in integrating mammalian reproduction, since it is involved in both behavioral and physiological responses to sex pheromones—the glandular secretions given off into the air by an animal.

It is of interest to us that this vomeronasal, or Jacobson's, organ exists in a passing phase in human fetal development, although it is merely vestigial in an adult, again emphasizing the importance of the sense of smell in our own evolutionary development.

There is significance to be drawn from the fact that the two areas in which the olfactory sense still plays a dominant role in us are the areas of eating and of sex.

In all species the most important business of their existence is efficiency in self-preservation and self-perpetuation. Every aspect of their structure, every function of their bodies, every feature of their habits and behavior, is honed by selection to further their effective ability to preserve and perpetuate not only themselves as individuals but above all their breeding group, and thus their species as a whole.

The physical and behavioral traits that serve this most elementary purpose are the most deeply embedded of any of a species' characteristics. That we ourselves, today, are still conscious of olfactory influences in whetting our appetites and in facilitating the fulfillment of our sex drive is some indication of the absolutely basic nature of the sense of smell in our remote evolutionary past.

Confirmation of such facts is often found in the oddest bits of information. For instance, the medical personnel of hospitals have often considered it a slightly humorous quirk that elderly patients, their faculties degenerating, frequently express sexual thoughts or show sexual interest. Cases of sexual jealousy in octogenarians—sometimes with incapacitated women as an object of it—have frequently been reported.

When we remember that any degenerative process begins with the loss of the most recently acquired function, while the oldest and once most vital remains to the end, this apparent quirk of the aged becomes understandable. Of their senses, hearing and sight are often the first to be impaired. But the sense of smell remains undiminished. This sense, unchecked and unmodified by the evidence of sight and hearing, brings the information it was evolved to monitor and it brings it in its own terms.

The sense of smell brings news of the bodies of surrounding creatures. The elderly person in whom sight and hearing are diminished is open to the language of his olfactory sense, and his remaining logic and reason can only fight a losing battle against this most ancient messenger.

We have seen many such cases, but one illustrates it dramatically. An eighty-six-year-old man made his wife's life a misery with his angry jealousy. There was nothing she could say that would reassure him of her fidelity. He constantly sniffed around, almost like a tracking dog, asserting he could detect telltale odors. His wife was a seventy-seven-year-old cripple, confined to a wheelchair!

Hoffer and Osmond noticed many similar "hallucinations" in their observations of schizophrenics. They were convinced that these events were in fact not hallucinations at all, but could be accounted for as a result of an increased acuity in the perception of odors.

The potential for gathering information via the nose was a part of our past and remains with us even though it has been overlaid by the dominance of the sight brain and the neocortical layer that endows us with reason. Had the evolution of our own species taken but a slightly different course, there is a possibility that today our reason might have been guided by olfactory rather than by visual principles. And so the speculation that somewhere in space such a race indeed exists is far from being an unlikely fantasy.

Chemical messengers that we call hormones carry out vital integrating functions *within* any animal organism. Pheromones perform similar coordinating functions *externally,* that is, between the individuals of the social group of which the animal is a part. When a female comes into estrus, for example, pheromones not only convey this message but also release the appropriate responses in the rest of the community. The pheromones, in fact, perform the external function of mediating social relationships and order. For that reason they have sometimes been called *ectohormones.*

There is hardly an animal society that does not in some way use a version of this scent language as its *dialect*—and we say dialect advisedly, because it is designed primarily to be understood only by the population that uses it. Closely related populations are often differentiated by the different odor languages they use; in that way, one population of, say, ants, may instantly recognize one of its own from a stranger or an "enemy within the gates."

While we humans are caught up with the concept of language as a *verbal* expression, there is absolutely no reason for a language to be confined to vocal signals. The possibilities of scent are at least as subtle as words, although they convey their meaning on a different channel and to a different destination. Traveling directly to the emotional center of the brain, they activate immediate responses, while words travel to the reasoning centers of the brain, which must reinterpret them and relay instructions to the appropriate executive centers. Scent language provides a direct line, whereas verbal language must pass through relay stations.

The chief function of the scent language, as of our own verbal one, is to regulate relationships between individuals and to aid in the ordering of the social group. Mating readiness and announcements discriminating friend from foe, information about the location of food sources, property rights—including the group's territorial boundaries and the individual's private right to a nesting site or home base—all are established by means of this language.

If these matters were not communicated and understood in a way that settled them unambiguously among the members of any society, constant infighting would threaten its order and it could not long survive. Chemical substances are at least as well suited as printed posters to marking boundaries, private properties, and as trail markers, and perhaps even better in some ways, in that they are readily apparent in darkness or when other objects might block out the view of a poster.

Deer, for example, have scent glands between their antlers and under the hair of their forelocks. Minute traces of this scent left on the trees and bushes of their range are just as effective as a fence or a KEEP OUT notice in indicating to other deer that it is private property. Bears reinforce their odor signals with visual ones by leaving scratch marks as high as they can reach on tree trunks; these function as calling cards that unmistakably indicate their size and strength. Almost every mammalian species has its own method of doing (or saying) the same thing.

Chemical messages play an even more significant role

among the social insects. The scent particles they use as a vocabulary represent a language no less flexible than our verbal one. Insects are able to use it, for instance, for a precise direction to a source of food, regulating the care of the next generation, recognition of group membership, the control of caste function, giving alarm signals, and every other social necessity.

Fire ants, when they go foraging for food, mark a trail of pheromones by extending their sting and touching the ground at intervals. Any worker coming across the trail by chance immediately recognizes the pheromone and follows its direction until it comes to the source of food. The amounts of pheromone deposited along the trail are modulated according to the quality or desirability of the found food as well as according to its quantity, and this provides a wonderfully efficient control over the distribution of the workers. There is no possibility of the worker being misled, because the trail-laying ant marks its path on the *return* journey only if it has found food.

All the ants arriving at the scene later eject pheromones in the same way as its discoverers if the food supply is still plentiful; in this way, they reinforce the pheromone concentration along the trail—which then becomes a powerful attraction to all the fire ants in the vicinity. When the food is exhausted, the then nonreinforced trail fades rather rapidly. (One of the various species of fire ants, *solenopsis*, follows gradients of concentration of carbon dioxide, possibly the simplest pheromone, to reach its fellows.)

When a threat to a colony of fire ants arises, a set of signals reverberates among the members, preparing them for whatever action is appropriate. Highly specific, these calls to action may be signals to attack an intruder; to call for additional help in repairing a breach in the wall of the nest; or for help to overcome a prey too large for a single ant to handle.

Fire ants receiving these messages display characteristic behavior: they raise their antennae and sweep them around in a semicircle as though they were radar aerials, and then they run in the direction of the disturbance. The alarm scent is broadcast repeatedly, like the intermittent wailing of a siren, to call for a continuous stream of workers; otherwise the nearby ants will continue their usual activities without interruption.

The alarm substances tell the ants not only the location and nature of the trouble, but also send them into a state of high excitement, which in the case of an attack becomes frenzied. The number of workers to leave the nest is signaled by the amount of pheromone—the alarm substance—emitted.

The pheromone language also announces the caste of individual nestmates as well as identifying colony members in general. This enables the ants to treat each other appropriately—to serve the queen or to tend the larvae—and to discriminate among life stages (we would say, they can tell their age), so that eggs, larvae, and pupae can be segregated into separate nurseries where they are tended according to the needs of their state. Ants also keep their nests meticulously clean. Feces are removed; dead ants are transported outside it onto a pile. Oleic acid odor, a by-product of decomposition, is the messenger that releases this behavior.

The chemical language of pheromones not only conveys information; it also promotes a suitable response. In some instances, pheromones not only trigger correct behavior but also govern anatomical development and the final structure of a creature's body. A termite larva that could develop into a nurse, a queen, a soldier, or a worker, receiving a scent message that some of the ranks are filled, then develops the anatomy of another caste.

These complex language systems, wonderful and sophisticated as they are, have come into being by evolutionary selection. An organ, or a behavioral habit, that works well—and all it needs is the slightest edge over similar ones—will in time prevail. Ultimately the need of the organism vis-à-vis its environment determines which of the possible ways it might develop.

As we have seen, our own evolutionary history goes a long way toward explaining why it is that when we smell a violet we get a *feeling* of joy (or whatever the feeling is), but we don't get an *intellectual* concept. For the Olfaxes this would be reversed. Having words to define concrete reality through smells, and brains developed to form abstract notions based on that information, all the evidence of their other senses would be meaningful to them only in those terms.

Assuming that their sense of sight is about as rudimentary as ours of smell, they will have no words to define fine differences of appearance. As we can say of a thing that it has a pleasing or an unpleasant smell, that it smells a little like an orange, a stable, or a fishmarket (which is describing its smell in visual terms), so the Olfaxes will only be able to say of a person's appearance that he or she looks good, and that the person looks a little like a horse, a monkey, a giraffe, or whatever crude image that person suggests to his mind. The sight of the person will evoke a feeling in them just as a smell evokes that kind of response in us, but they will be as baffled in describing this feeling objectively as we are when we try to define an odor. But the person's odor will have real and precise meaning for them.

We now come to one of the most fascinating aspects of the sense of smell. To describe it we have briefly to become a little technical, but we ask our reader to bear with us, because its implications are immense and they have been little explored.

The olfactory organ is attached to the ethmoid bone at the top of the nasal cavity. The nerve bundles, unlike those of any other nerve, do not form a cable of all their fibers, but each passes separately through its own small perforation of the cribriform plate of the ethmoid bone–that is, each has its own separate tunnel straight to the brain (the olfactory bulb).

This is a most primitive arrangement, dating back to the earliest marine vertebrates. Why it should have remained so, we cannot be sure, but there is a possibility that the sense of smell was so well developed at such an early stage that there was no further pressure for its improvement. Any further refinement apparently took place in the brain itself, in its elaboration of the already sufficient scent information it received.

As the olfactory nerve bundles reach the bulb, they couple with formations called *glomeruli*, where the first junction with other nerve fibers occurs. There are some two thousand glomeruli in the human olfactory bulb, each one emitting a vast network of fibers that form a circular pathway (or, in technical terms, a reverberatory circuit). Like a regenerating radio receiver network, this arrangement allows for an amplification of the nerve's olfactory messages, and thus adds another magnitude to the olfactory sense.

This complex network within the brain has important implications. From the structure of the rest of the nerves in their course from the olfactory organ to the brain, it would seem that in earlier evolutionary stages they delivered their messages by point-to-point transmission. However, the final elaboration within the brain itself suggest a later-developed complexity to this sense that may convey *a spatial patterning analogous to the three-dimensional quality of vision*, perhaps like a hologram.

If this is the case, as it seems to be, then animals receiving olfactory messages from other creatures or objects in nature can perceive them "three-dimensionally" via the sense of smell, much as we perceive a three-dimensional visual image. For example, a dog approaching another animal receives scent messages. Because of our unfamiliarity with this type of sensory information, we assume that it receives a single, specific message: for instance, "deer."

In fact, however, we may assume from the construction of the dog's olfactory brain that it gets graded messages from different parts of the deer's body, probably forming a three-dimensional pattern in the dog's brain. The dog will be able to perceive and recognize even in darkness what kind of animal it is, its size, shape, and anatomical features. But the dog would know more than we do, even via sight. It could also discern from the deer's various glandular secretions whether the deer is hostile or afraid, male or female, in heat or other states of health or mood.

So far as we know, most animals interpret these kinds of odors only in their own species or in species with which they are closely involved, as in a predator-prey relationship. On the other hand, some examples of cross-species scent recognition have been known. For example, women in Yellowstone and

Glacier National Parks are warned not to sleep in the open or in tents when they are menstruating, because the bears that inhabit those areas have been known to attack and severely maul women in this condition.

The three-dimensional perception afforded by a keen olfactory sense—with the added dimension of recognition of emotions, moods, and various physical conditions—is a form of awareness that we can hardly imagine since we cannot apprehend it visually.

But imagine this sense at the disposal of a species of high intelligence, as we hypothesize the Olfaxes to be. Once they had familiarized themselves with our structure and psychophysiological responses, we should be open books to them; deception would be out of the question. Our moods and intentions would be broadcast before we expressed them, and sometimes even before we became aware of them ourselves. It would seem that if we ever come to a point of sending astronauts to visit the Olfaxes, we shall have to stress the necessity of the old-fashioned virtue of honesty!

If we think a little further about the nature of the olfactory bulb and the three-dimensional quality it imparts to the sense of smell, we find ourselves like Cortez, gazing out at a vast new and unexplored domain. An understanding of this "in-depth" quality of the olfactory brain offers a tantalizing glimpse of the unexplored land of our emotions, which are closely linked to the sense of smell.

We describe our emotions with words that have spatial connotations: "*wave* of pity," "*flow* of feeling," "*height* of ecstasy," "*depths* of despair," "*surge* of love," "a *welling up* of envy." Moreover, we think of emotions as waxing and waning, persisting or lingering, which implies a diffusion over time, again not unlike the similar lingering qualities of the sense of smell. Beyond the spatial quality and its extension into time, there are other words—*overwhelming, overpowering, intolerable, unbearable*—that we use to describe both emotions and smells.

These conceptual and verbal connections are firmly rooted

in the way our brain is constructed. If we compare an emotional experience with the single-dimensional quality of the products of our reason, the contrast emerges sharply. When we "know" something via our neocortex, it has a quality of precision. It is, or it is not. A calculation is correct or incorrect; an action is taken or not taken; we should proceed this way or that way. The considerable achievements of our reasoning brain have resulted from its logical, step-by-step processes, results piled upon accumulated results, inferences made and conclusions drawn, until vast edifices are built.

Nevertheless, every once in a while we catch an inkling of other pathways to accomplishment—flashes of intuition, leaps of imagination, the insight of genius—that arise from other parts of our minds than our reason.

The nature of this kind of mental process has puzzled philosophers and scientists alike over the ages. Ancient peoples attributed it to the voices of the gods. Today we give it a name—heuristic reasoning—but this does not explain it.

We cannot explain it for the same reason that we cannot define an odor—because it arises in a part of the mind over which reason has no jurisdiction. Perhaps it is one of those potentials that exist in us not fully realized, but that nevertheless offer us hints for the hypothetical construction of the intelligences of other worlds.

The
Olfaxes

THE DATE IS SOMETIME in the future, and we are approaching the planet of the Olfaxes. As we begin to penetrate their moisture-laden, densely clouded atmosphere, our radarscopes indicate a plain beyond a hilly region, and we are heading in that direction toward a soft landing.

Preliminary work undertaken before we set out has indicated that this planet's atmosphere, although different from our own, is capable of sustaining life and appears to contain sufficient oxygen to support us without harm to our metabolism.

As we open the hatches of our spacecraft and step out onto their soil, we find immediate confirmation of some of the many tests that were carried out before we left. The first thing we notice is that visibility is very limited. We can see fairly distinctly only within a range of about 50 meters; objects farther away dim out.

The dense cloud seems to drain all color from the landscape, as though the objects of it were seen at twilight. We make out the outlines of vegetation not too dissimilar from what is found in a tropical forest on our own planet. As our eyes accustom themselves to the dim light, the forms fill out and we see them as though in an underexposed black-and-white photograph, all in shades of gray. Here and there we discern what seem to be giant ferns and many forms of outsized mushrooms, although most of the vegetation is unfamiliar to us. In the absence of direct sunlight, the plant life here must have evolved a way of storing and utilizing energy with only reduced possibilities for the photosynthesis by which our vegetation exists.

We notice that the climate is warm and moist, as in a greenhouse. The cloud cover has produced the greenhouse effect by allowing heat waves, rather than light or ultraviolet rays, which have been deflected by the outer layers, to penetrate it.

At first sight it seems to us that all the chemical processes of life that on earth are dependent upon light (the production of some vitamins, for example, as well as photosynthesis) must here be dependent on the utilization of heat energy, and probably by means of different chemical and physical processes.

Fortunately we had taken into consideration before leaving earth the probability that the carbon dioxide content of the atmosphere of this planet would be higher than our own, and have equipped ourselves with filtering masks. While we can tolerate only a limited amount of this gas, the creatures here must have developed a way to accommodate to it. But some oxygen is available. We have to wear our masks only at intervals when the slightly higher level of carbon dioxide begins to make us breathe more heavily and then to feel drowsy. We assume that the plant life we see, although not able to photosynthesize carbon dioxide and water into starch in sufficient quantity, must have evolved enzymes that can do the same thing in the absence of sufficient light.

As we proceed, our portable radarscopes indicate that we are not far from what could be an inhabited center, and we direct ourselves toward it.

Soon we detect, looming out of the hazy shapes ahead of us, what seems to be a settlement of conical edifices of various sizes, some with flattened tops, some with excrescences circling them, and some set, mushroom-like, on sturdy central pedestals. Yet as we near this center, which looks as though intelligent creatures have built it for living or working purposes, we are surprised at first to see neither population nor traffic moving about in it.

As we pause, wondering which way to proceed, we suddenly see, rising from the ground in front of us, a circular structure which comes to a halt when it has risen some 3 meters. Its wall slides around, and about a dozen individuals emerge from it, whereupon it submerges again until its roof is level with the ground. This is our first glimpse of the Olfaxes.

Our first sensation is relief. Although we actually know better, we cannot help having had some trepidation about the kind of creatures we were going to find, and our imaginations had conjured up some weird and wonderful fantasies. But the Olfaxes turn out to be unalarming, and externally, at least, not so very different from ourselves. The first thing that strikes us is that their noses are very long. Otherwise, like us, they have four limbs, two of which are adapted for standing and locomotion and two for holding and manipulating—we see them carrying objects much as we do.

However, the Olfaxes' limbs are rather short in proportion to their bodies as compared with ours (the sort of proportions a dachshund might present if it were an upright creature), so that natural locomotion is obviously slower and more strenuous for them. A little later on, we find that they have compensated for this disadvantage by using a great variety of transport, and they use it almost to the exclusion of walking. This is why the plazas of their "town" were so empty of strollers.

It soon becomes apparent that the large and long proboscis—nose—besides being a highly refined olfactory organ, is also an additional prehensile one that aids their short arms as a third upper limb. This proboscis grasps by suction rather than by clasping, and has given the Olfaxes a great deal of dexterity on two accounts: first, because of the availability of three manipulative limbs, and second, because one of these limbs, operating on a different principle from the other two, has stimulated a portion of their brains that is in closest contact with their most highly developed sense, the olfactory. This clearly must have had a circular effect, the function of the proboscis stimulating a part of the brain, along with the manual skill, while the stimulated brain part, hypertrophied, acts as a spur to more varied and more efficient use of the proboscis.

That this has eventually led to a technology based on the mechanical principles involved in suction is one of the first aspects of Olfax life that become apparent to us.

Two of the Olfaxes that emerge from the rising turret gesture to us. They make sounds that must be part of their verbal communication and that we cannot understand, but their bodily gestures and expressions unmistakably invite us to accompany them the next time the small turrent rises, disgorging some more Olfaxes. As it empties we enter and descend in it with them, and find ourselves at a hub of what is a busy transportation system.

We see a great variety of vehicles, at this point based on vacuum-tube and suction principles. Later we learn that they also use wheels, steam, and electric power for many purposes, but here, in the first of their subterranean roadways we discover, the vacuum-tube principle is in operation. Through it, large vehicles serving greater numbers, and small ones accommodating any number from a single individual up to about eight, are propelled.

We are impressed with how the technology of an intelligent species inevitably derives not only from the logical ability of the brain but also from the inspiration gained from the physical use of its own body. With us, for instance, our mathematical calculations are made in terms of a system of numbers based on ten, the number of our fingers, our first counting instruments. Indeed, our language encodes the connection, in that we use the word *digit* to designate both a finger and a number. The Olfaxes, accustomed by their bodily structure to use suction equally with grasping or pushing and pulling to grip and move objects, naturally incorporate this method in their technology.

As we now move among a greater number of Olfaxes, we notice that the clothing they wear does not seem to be designed for protection. In keeping with their warm, moist atmosphere, it is lightweight, almost flimsy. We see similar garments on large numbers of individuals, like club uniforms, and distinguished more by the number and arrangement of small pockets upon them than by style or color. We learn later that the pockets are in fact perfume sachets and that the clothing adds to the Olfaxes' personal identifying odors information about the ceremonial rank they hold in their social communities, very much as uniforms and titles immediately convey this kind of information among us.

Our journey in the vacuum tube is brief. We leave with our guides and again enter one of the elevator-exits. This one, however, does not rise into the open, but directly into the pedestal base of a mushroom-shaped structure similar to those we had glimpsed at our first view of the Olfax settlement.

This one seems to us to be a ceremonial hall because of its size and sparse furnishing, and indeed it proves to be so, even though we later discover that the interior arrangements of all Olfax habitations are strictly functional and without ornamentation. As we enter we become aware of an unfamiliar odor that to us has a slightly floral or spicy association, something like the subdued but subtle smell of nasturtiums.

What we are writing now is an account as though in retrospect. On arrival many of the things that later became clear to us were puzzling. That we were immediately greeted in a friendly manner was one of them. Later on we understood that hostility, like fear, has an odor of its own and since we gave off no such odor it was assumed that we came either in friendship or curiosity, but that in either case they knew even from a distance, before they met us, that we had no ill intent toward them.

It soon became clear to the Olfaxes who were delegated to receive and interview us ("interview," of course, is our word, not theirs, which could more closely be rendered as "smell us out") that a period of initiation would be necessary on both sides before we could have any useful exchange of ideas. This was borne in on them as well as on us almost at once when, according to the program worked out before we left earth, we tried to show them our place of origin by offering them celestial maps and charts. They reciprocated by placing pages below our noses, but we could not discern anything recognizable upon them.

Both sides then realized that we would have to master something of the other's verbal communication, although later on, even when we had done this and learned to comprehend each other's words, we still found it extremely difficult and at times impossible to understand each other's concepts since we cannot precisely define an odor, let alone build an idea on it, while they are equally unable to define exactly a visual impression.

This point came out right away when we tried to exchange maps and charts. They have developed olfactory hieroglyphs using microscopic globules of odorous substances (of the type that have been used by some advertisers among us, where one can rub a printed image and obtain its specific smell), much as we use printer's ink to convey printed, and therefore visual, abstractions of our spoken words. Their records were almost invisible and totally meaningless to us.

Those we first saw had been made by injecting these globules into absorbent surfaces made from a suitable vegetable material reminiscent of our paper. Later we saw other types of material made from skins and from chemicals that were analogous with our parchment and photographic film. They had never infused their odor globules into stone as our earliest writers incised their visual messages, but some of their forerunners had found a way of using impacted sand and clay before arriving later at the production of more convenient substances.

Thus they had found ways of fixing their olfactory messages upon such surfaces as we have found ways of fixing visual ones on our printed pages. When we eventually entered one of their libraries we recognized it for what it was by the stacking of sheets, scrolls, and rolls of their volumes, which had been stored with insulating material separating and preserving them. Even though their contents were imperceptible to us, beyond our awareness of a generalized confusion of, to us, unidentifiable odors, by then we knew them to be a very close equivalent of our books. As their technology had progressed they had also found ways of transducing their olfactory records into magnetic signals that were more easily stored.

Basically our own visual world of lines and curves is abstracted in our writing by using uniform lines and curves arbitrarily to symbolize sounds. Thus we render speech visually. What the Olfaxes had done was to use odors in the same way, identifying certain fundamental odor elements with certain sounds or words, and had thus built up a system of olfactory ideographs that served them as effectively as our written words serve us. An Olfax printer dips into vials of standardized odorous chemicals and spaces these either manually or mechanically on a page, as our printer selects letters from a font of type and arranges them to conform with our written code.

By this means, we were eventually able to begin to understand each other's words. They used their magnetic coils, transducing them into visual signals for our benefit, and then back into olfactory ones for their own. Thus we got a visual code approximately like a voiceprint, while they got their odorous hieroglyphs from the same magnetic coils.

At first we noticed that they deliberately sniffed at objects before uttering a word, and then we came to understand that this was their equivalent of pointing with a finger, which is a visual cue. In time they understood when we pointed, and so we came to learn each other's words for objects, and after that to master a working knowledge of our languages.

A comprehension of their syntax was a great hurdle for us, and ours for them. A sense of the past, present, and future was utterly different in their thinking from ours. Due to the lingering nature of their predominant sense perception, much of what we would consider to be final, or past, was included in and modified both their language and their ideas about the present.

There was also an initial difficulty in quite casual remarks. Their thoughts about what is significant and what trivial are quite different from ours. Once one of them made a comment about a leaf that was completely unremarkable to us, and it took us a long time to realize that the Olfax was making a reference to its singular odor (by the way, imperceptible to us) in a metaphorical way. At another time one of us said how aesthetically pleasing we found the stark functionality of their interior architecture, and this was a thought that could only be conveyed to them with a great deal of mutual effort.

Their architecture, both interior and exterior, made the strongest impression on us. It was devoid of spires, minarets, domes, external or internal decorations, whether of sculpture, painting, mosaics, tapestries, murals, gleaming chandeliers or silver candelabras, ornate wrought-iron work, enamels, richly textured or colored fabrics, or beautifully crafted bibelots, and all the things that go into our houses and public places to please our eyes. On the other hand, homes that were luxurious in their terms, although severely functional in design, had walls fabricated of aromatic woods like our cedar or sandalwood, but in many more varieties and treated to render their odors almost everlasting.

Their artists produced not paintings or sculptures for these homes, but exquisite concoctions of subtle perfumes of infinite variety. They varied their effects even further by arranging and rearranging the placement of these perfumes in their rooms: in nooks; high or low on walls; at various distances from each other; and in different positions relative to each other, changing the effect of each on the resulting blends, so that they could mingle and intermingle changeably, creating a symphony of odors as a background for gracious life.

In less all-pervasive designs we sometimes saw walls inlaid in small areas that at first we took to be like paintings, but that turned out to be a medley of odors. Their Raphaels and Michelangelos had selected, blended, and arranged perfumes as originally and as perceptively as the Japanese writer of haiku selects his words or as the Chinese artist isolates or combines images of bamboo stems to convey a sense of beauty and to abstract from these elements the wonder of all nature. Moreover, the Olfax artist, by associating perfumes that have a connotation of fields, individuals, rituals, or edifices, within a framed area could produce in his audience by olfactory means a response similar to ours when we see painted lines and colors on a canvas that combine aesthetically for us and also produce a visual image of the things they represent.

Another aspect of their architecture that was of interest to us was that they usually used above-head areas, similar to skylights, when they wished light to percolate, since their equivalent of windows were not for the sake of a view but only for a modicum of light. But their walls did contain vents to permit the flow of air and the reception of odors from outside.

One of the things we soon became aware of was that the life of the Olfaxes was very much centered around extended family groups. What we would call their town or village centers were not very large. The populations were more evenly distributed over the whole area of their planet than ours are. We thought that this had something to do with sensitivity to odors, but it probably also had economic and historic reasons that we did not at first perceive. That they did not have stark mountain ranges nor vast oceans but rather gentler hills and many small lakes, so that their whole planet surface was habitable, was also surely a factor.

We were soon impressed by the extremely high intelligence of our hosts, and also with the serene quality of their family relationships. Apparently forewarned of mood changes by pheromone messages, they found it easier to adapt to each other's emotions before they became extreme or frustrated.

Much of their entertainment was within their homes. Clearly, any intelligent species must entertain itself, since any evolved organ has to be exercised in order to remain in good condition, and the brain more than any other. As we ourselves know, our brains punish us with the extreme discomfort we call boredom if we do not exercise them sufficiently.

We noticed that each family group of Olfaxes had an instrument that looked to us something like a prayer wheel, or spindle, in the central area that was a feature of their homes. These were loaded with odors associated with actions and scenes in the natural world, and when rotated they seemed to have the same content for an Olfax as a silent home movie would have for us. We sometimes also saw a more elaborate version of this wheel, with a recorded sound element, so that this told a story for them via smell and sound, as our movies do for us via sight and sound.

They also had an instrument we came to equate with television. So far as we could understand, it seems that in earlier days odors were piped through into each habitation for public announcements and news as well as for pleasurable entertainment, but that as their technology became more sophisticated their verbal language as well as the odor images were converted into electric and shortwave pulses and transmitted in a way similar to the transmission of sights and sounds over our television systems, then unscrambled and converted back to odor and sound messages by the receivers in their individual homes. In fact it was a system of teleolfaction.

It was quite a while before we were able to take for granted the fact that for the Olfaxes odor played a role as encompassing as vision for us, and a role of the same nature. For them odor was not an adjunct of an object; it presented them with, if we may call it so, an image of that object. It was their *experience* of it. It was not that they were without sight any more than we are without a sense of smell, but for them the *appearance* of any thing was an adjunct of the complete comprehension of it they had through its smell.

In one way their language, codified from abstractions of their odor perception, was less vital to them than ours is to us. Since in the dark and in obstructed places we cannot see, our most urgent public announcements and alerting signals are relayed to us by loudspeaker systems and other *sound* signals like sirens, whistles, bugles, horns, church bells, and, in some parts of our world, drums.

This was not necessary for the Olfaxes, since their odor signals could be broadcast direct without the intermediary of language, and therefore all their emergency and warning alerts were made this way. Actually they were more effective than our beacon-light or red-light signals, because we have to be looking in the direction of a warning light to see it, but an odor, like sound, is pervasive and makes one immediately aware of it.

Quite soon, even before we were conversant with their verbal language, we realized that the extended family groups in which they lived were organized differently from our human ones. This unit of their social organization was more of a compact clan type of grouping. There was not a great deal of difference between the superficial appearance of their males and females, either in their external bodily form or in the garments they wore. The Olfax clan groups were rigidly hierarchical, with roles determined more by rank than by sex. Among them the ranking male and female took responsibility for the well-being of their family group, aided by next-ranking family members of both sexes, and obeyed by the subordinate ones. The ranking female Olfax abdicated the economic and protective leadership role she held in tandem with her mate only for a minimal period when her offspring was due, and she then returned to the pleasures of her duties after that short interlude, leaving her young ones to the communal care of the rest of her family group.

When writing the words "minimum period" we were on the point of saying "about a month," but this reminded us that the planet of the Olfaxes has not one but two moons, and so the lunar influences on the physiology of all the creatures living there is dual, and at times overlaps in a way that is extremely complex and would have to be studied very carefully, especially as to the way their metabolism is affected, for full understanding.

A side effect of the Olfax "clan" or extended-family system was that it eliminated the necessity for large laboring classes, since each group was to a large extent self-sustaining. Within the group, roles were rigidly defined, but these changed with the age of the individual members and the capacities they displayed as they grew to maturity. It was the responsibility of the leaders of the groups to assign tasks and to ensure the group's care and viability. But if a member desired a change of occupation this was usually accomplished by a period of apprenticeship training with others qualified in the desired field, so long as a willing substitute could be found for the task being given up. However, there were some clan groups that traditionally occupied themselves with mining and technology, and it was these who were responsible for the transportation systems, the vehicles, the "printing" technology, and the building and domestic arts (including the utilization of electricity and natural gas) that we had already seen.

Not only was the family pattern hierarchical but, as we soon learned, the whole of their society reflected this. We found it interesting to compare the convention basic to their hierarchy formation with what exists among us. On earth, of course, strength, martial ability, sexuality—whether masculine or feminine—and competitive edge in special fields underlie leadership quality. The presence of hierarchy in mankind usually implies competition without letup from earliest childhood and throughout life—and even earlier, right from birth, in other animals on earth.

Among us there have been few peoples, and at that only for brief periods in their history, where qualities of intellect have endowed their bearers with the privileges of rank. Ancient Greece at the time of Pericles was one of them. There, those whose mental abilities and curiosity about the nature of the universe prompted them to devote their time and thought to the pursuit of natural philosophy became the respected leaders of their society. The arenas of their contests were not battles nor formalized jousts, but the symposia, the debates in which they expounded their ideas. This form of social contest was possible because their material needs and wants were provided by a slave caste.

One could find analogies among some of the birds of our planet that attain dominance by the formal display of plumage or of ornamental bowers that have no practical use. For these birds the necessities of life are all around them for the taking and so it is not necessary for them to prove efficiency in providing the means of life to obtain mates and produce young.

In human societies where religious beliefs have fostered the rise to dominance of priestly chiefs, we have also seen qualities of intellect superseding qualities of strength in the allocation of prestige and leadership position.

The situation was similar among the Olfaxes. The nature of their climate and material environment made the matter of sustenance merely a question of organizing its collection, and so the administratively and intellectually endowed came to be recognized as their natural leaders. They were not a warlike race and any aggressiveness in their character makeup was turned toward the exercise of the mind.

This had not always been so. Advanced technology, wherever one finds it, is associated with an aggressive, expansive, and hostile population, and there was a time in their earlier history when the Olfaxes were much more technologically oriented than they were at the time of our visit. Many of their technical amenities were legacies from those days and had been carried on like traditional crafts without too much innovation ever since. For several thousand years now, they have lived in a stable type of state.

This very pleasant state of affairs had not come about by chance. At that earlier time in their history their survival had depended upon endless contest, caused by the overproduction of offspring, in exactly the same way that natural selection operates among us. Wherever the population exceeds the ability of habitats to support it, social dominance has to be sustained by proven ability to provide the means of life, and the less efficient fall by the wayside.

But the forebears of the Olfaxes thousands of years before we came to their planet had recognized and faced this problem. They had taken effective measures to thin out the density of their populations and they had devised means to control fertility, so that by the time we met them the Olfaxes were enjoying the benefits of rational planning that had eliminated overcrowding and regulated social order by the exercise of intellect. All this had been achieved by olfactory means.

Of all Olfax arts, the culinary was the most highly esteemed and the most exquisite. Even more than we are affected by and attracted to the sight and smell of food as well as its taste, the odors of Olfax food, beyond its taste, were a source of a highly complex etiquette in entertaining. The role of food in their lives was central, because it did a great deal more for them than provide sustenance. Outstanding Olfax hosts and hostesses conveyed so much meaning by the odors of foods that conversation at table was barely necessary—the food would almost speak for itself, either sexily, admiringly, formally, cosily, or analytically. Their meals fed the appetite, the emotions, and the intelligence of their guests simultaneously.

We felt like country cousins in our very limited ability to experience those occasions as fully as they did. We could not help but wonder what we could have done with the resources we possess had the Olfaxes been visitors to our planet rather than we to theirs. Could our kitchens have been our Rosetta stone to unscramble Olfax ideographs? Trial and error might have found a way to them of ideas, at first simple—like, "You are welcome," or, "We like you"—and later more complex.

When we talked of this among ourselves we decided that courtesy might dictate that initially we supply them with surgical masks, even gas masks, until they became able to accustom themselves to the, to them, surely stunningly wild jumble of odors and associated information—or rather, since it is haphazard, the misinformation—that would greet them.

But to return to the Olfax table, one of its delights was the service of food on thick leaves that were afterward discarded, or on platters of wood that were not used many times before being buried in the soil and used no more. In this way the Olfaxes avoided overutilization of their vegetable resources, since their soil was constantly enriched with refuse compatible with its ability to break it down and recycle it into plant life, and never contaminated with semi-indestructible artifacts.

This rich soil produced the vegetable foods of high quality that formed the staples of their diet. They did also eat animal food; most of the family groups possessed some animals domesticated for the purpose. The greatest proportion of their animal species were burrowers and lived in the soil below its surface.

Naturally there were many forms of animal life on the planet. No higher forms, and certainly no intelligent species, perfectly adapted to the physical circumstances of its environment, can possibly exist unless lower forms are present. From what else could it have evolved? The only other possibility would be for it to have arrived, fully developed, from some other part of the universe, and this was not the case with the Olfaxes. There, the dim light aboveground on the one hand and the very rich, fertile soil on the other, seem to have been factors in promoting greater numbers of soil-living species than aboveground-living ones.

There were many more varieties of wormlike creatures than exist on earth, and also more species that one might equate (although their appearances were totally different) with our moles, rabbits, shrews, woodchucks, gophers, and other creatures that live in burrowed-out colonies below the ground. Some of their larger creatures, too, lived below ground. Whereas among us foxes, wolves, or badgers excavate lairs only to rear their young, their animals of that size emerged aboveground only when searching for vegetable food other than roots.

There were not many larger animals. Animal life is inclined to adjust its size, as everything else, to its ecological niche, and the planet of the Olfaxes is smaller than ours. (The Olfaxes themselves are not a large race, the tallest of them being between four and five feet high.) There was also animal life in their waters, and there again the species were small. The waters of the planet had accumulated in many small lakes, but there were no great oceans to accommodate giant marine life.

Insect life existed there too, but by far not in such abundance as with us. In the first place, many of their plant species propagated themselves by vegetative means, their roots running underground and producing new shoots along them, or by spores (like the ferns and mushrooms) that are relatively independent of insects. Their flowers, unlike ours, are not attractively colored. Since all colors were in any case blanched out to grayish, brownish, or vaguely gray-greenish tones by the perpetual overcast, bright floral colorings and markings would have served no adaptive purpose and, indeed, had not evolved.

Instead, an enchantment of odors of infinite variety lured the few insect types that did exist to aid their cross-pollination. In the absence of sufficient insects, though, the plants largely depended upon their own structural mechanisms to achieve this. Bird life was absent. Seeds were carried by wind, scattered by dehiscent fruits, or passed through the bodies of animals in order to return to the earth and root themselves.

The high intelligence of the Olfaxes themselves arose, as it had to, from the circumstances of their evolution. Their species stemmed from an earlier one that lived largely on vegetation and partially on grubs and other small creatures that inhabited the soil. Their short limbs were an advantage to this earlier form, which possessed no offensive organs like tusks, fangs, or claws, but depended upon its ability to remain hidden among the plants that were its source of food, for protection from the larger, better-armed predators. The larger predators, which at that period roamed their planet, shared their habitat, and preyed upon them, seem to have been the dominant form of life then; to judge from the successful "hiding defense," we have to assume that they were more visually oriented than the creatures extant on the Olfax planet now. As was the case with our dinosaurs, circumstances—perhaps climatic—contributed to their extinction, and predominantly olfactory species took over.

Early in the protohistory of the Olfaxes, those individuals that possessed a particularly keen sense of smell had a survival advantage over the rest, since they could take cover sooner and avoid annihilation. In the course of time a variant arose that possessed not only this keen olfactory perception of the distant presence of other organisms, but also a means of modifying the odors given off by its own body. It could, so to speak, turn up its olfactory emanations for the benefit of its own group (for communication, identification, and similar purposes), but reduce them or even change them slightly to throw off possible predators—much as we raise or lower our voices at will, or alter our tone for communication, to preserve our place of hiding, or to announce our presence or to raise an alarm.

This ability became the hallmark of the species' developing intelligence, and became its ultimate weapon, a weapon that was not merely a compensation for its comparative physical weakness, but an overcompensation, reaching a new dimension.

Like our young, and the young of any other intelligent creature, their young needed a long learning period, promoted by slow growth and prolonged development. Inevitably, this fostered attachment behavior between the young and the adults that cared for them, which in turn demanded moral values similar to our own, based on giving care to those who need it.

Wherever intelligence of a high order exists, it has to have evolved along with a prolonged childhood period for learning. This extends into evolutionary selection for adults with a bias toward care-giving paralleled with infants with innate attachment behavior. In time this caring/attaching propensity brings its own problems in the preservation of the unfit, of youthful features in adult behavior like curiosity, manipulation, and playfulness. Intelligence is concomitant with prolonged youthfulness and forms a self-perpetuating feedback in which each of these characteristics reinforces the other—and it is likely to appear in this way wherever in the universe either of them arises.

In line with their characteristic curiosity, modern Olfaxes are in no way behind our own scientists in their investigations about the nature of the universe. They have found ways of registering the various forms of energy that penetrate their atmosphere from outer space by methods that allowed those energies to interact with chemicals and give off odors. Thus an Olfax scientist, through the use of olfactometers, can reach accurate conclusions about the nature of their sun in a similar way that ours draw conclusions from their use of telescopes.

For example, ultraviolet rays from the sun do not penetrate their cloud cover, and the Olfaxes therefore have no natural experience of them. Nevertheless, they produce these and other energy waves by heating various substances in their laboratories. When ultraviolet rays interact with oxygen they form ozone, which even sharply sighted creatures do not see, but smell, and therefore become aware of its presence. The Olfaxes therefore have no difficulty at all in deducing the existence of ultraviolet radiation from the presence of ozone.

Similarly, the interactions of other energy waves with any substance produce subtle odor effects that escape us but give clues to the Olfaxes. In this way they readily recognize X rays and some of the radioactive emanations. They can perceive, if indirectly, everything that we can perceive, by conversion into their olfactory mode, as we convert such invisible energies into visual or acoustic codes.

Since the Olfaxes do have a sight sense just as we do have a sense of smell, one could ask why they do not transduce their laboratory findings into graphs or such modalities as our radar screens, showing blips to indicate invisible objects, but we discovered that such charts and signals were as meaningless to them as their codified odor symbols were to us, in spite of our ability to discern some of them. On the other hand, they did sometimes use acoustic signals, of the type of our Geiger counters, for similar purposes. Indeed, the acoustic sense was the only meeting ground we had for communication, since both they and we could convey meaning via sound.

Arising from this, the basis and structure of their mathematics is radically different from ours, derived as it is from minds that perceive the world in terms of odors. For instance, when we count from 1 to 12 (the Olfaxes have six digits on each hand and their mathematics are based on a duodecimal system), we have a concept of twelve separate integers, each with a specific numerical connotation. For the Olfaxes, with their sense of the diffusion of matter, their numbers represent gradients between each integer. The number 1 represents a field extending from 1 to 2 , and so on along the line. As a result, their mathematical calculations are expressed in symbols of probability and utilize the concept of statistical averages far more than the absolutes of our digital form of calculation.

Another outcome of their sense of the diffusion of, to us, objective things is apparent in their sense of time. We have mentioned that to them much of the past flows into the present and coexists with it, so that their thinking patterns and their language are based on different premises from ours, especially in this matter of what is past, what is present, and what is future.

This conceptual and verbal indefinition is carried over into naming the hours of the day. It was not their practice to use any equivalent of watches or clocks to pin down the precise time. For them the day was divided into periods that carried names approximately equivalent to: premorning, very early morning, early morning, midmorning, late morning, midday, almost late day, late day, very late day, and so on. They could be more precise if necessary, and they had words that subdivided these periods into smaller time spans, and still smaller ones for scientific purposes, but it was not their habit to use the more precise terms in their ordinary lives.

Of course, the same dim light that was the reason for the supremacy of their olfactory sense also played a part in this, since there was not a great difference in the degree of light during day or night, nor was there a sharp demarcation at dawn and sunset. In the early days of our contact with them this was a source of difficulty with us. They could never understand why we wished to be so precise in the allocation of our time. An exact appointment was entirely outside their way of thinking.

Their mathematics and their sense of time were but natural extensions of the intrinsic nature of the olfactory brain. This can best be illustrated in our own world by our attitudes when we refer to our emotions, which, as we have noted, are mediated by the same part of the brain as the sense of smell. If we ask a person, "Are you happy?" he will stop to think about it and answer imprecisely that yes, on the whole, he is. But if we then ask him, "How happy?" he will be totally unable to give us a quantitative definition of the degree of his happiness. So it is with the Olfaxes, and it presents a considerable handicap to communication between species of our type and theirs.

The political life of the Olfaxes was simplified by the fact that nation states did not exist there. Since the planet is relatively small and lacks insuperable geographical barriers, and since the population is rather evenly distributed over it, circumstances had not lent themselves to the development of such political entities. Decisions for the whole community were made by an assembly composed of the couples that headed each clan unit. From among them the members of the assembly nominated by consensus individuals expert in certain skills to organize such technical and general services as were necessary. Thus a kind of technocratic elite superimposed upon a responsible autocracy governed the population.

It was not a pyramidal hierarchy, but structured rather like a pyramid flattened or rounded off at the top, and it struck us that this type of political organization must have affected their feelings about desirable design, for it was reflected in the predominantly mushroom shape of their dwellings.

The supremacy of olfactory perception among these intelligent beings had certainly been responsible to a large extent for the peacefulness and good order of their society. Where emotions are perceived before they become violent they can be accommodated and fighting avoided. The Olfaxes possessed no offensive weaponry. They did not need it even for hunting, since theirs was a gathering economy and the animal food they used was domesticated. Their fertile soil made life easy for them in this respect.

But this state of affairs had its negative aspects also. Almost like our social insects whose extremely complex social groups are also governed by olfactory communication, Olfax society had stagnated. It had remained stable for so long that new ideas were felt to be disruptive rather than challenging and were usually suppressed. All their greatest achievements had taken place in the distant past.

Original ideas produced by their intelligent minds were diffused and diluted into their favorite entertainments—discussions, mathematical problems, puzzles, the refined appreciation of odor messages, and intellectual games. Physical games had no place where there was no need to train bodies and reflexes for fighting, and technology beyond that which facilitated their domestic ease had no incentive without the demands of warfare or intense political competition to initiate it. With no oceans to explore and no birds to emulate in striving for flight, these areas of contest with the elements were also nonexistent.

We had been struck by the peaceable way in which the Olfaxes had received us, their curiosity stimulated but with no apprehension or thought of possible hostility from us. We could not help thinking that should some other invaders, less peaceably disposed than we, ever find their way to the planet of the Olfaxes, this peaceful and in many ways ideal race could not survive for long.

FOUR

The Polarized
Light Sense

LET US POSTULATE A creature that leaves its home base in search of its daily food. It zigzags, loops, circles, goes forward, and doubles back over the range of its search in a landscape with so many and such diverse features that a human being would without doubt get lost in it. Having eventually located and collected some food at considerable time and distance from its starting point, this creature then has no need to retrace its path, but turns around and without any instrumental aid—on the basis of its directional sense alone—finds its way directly back home in an unerringly straight line.

To duplicate this feat a human would need a compass, a stopwatch, an integral vector calculus, a polarimeter, and perhaps a portable computer. We should find such an accomplishment impossible without these aids. Yet on earth there are creatures that have this ability: honeybees, for example, and ants, with their incredibly small brains—smaller than a grain of millet seed—do it every day of their lives.

Let us observe a bee or an ant and follow it step by step to see exactly what it does. As it leaves its hive or hill, it perceives the angles of the path it takes relative to the sun. To do this it does not have to observe its direction and consciously make calculations, as we would. It is possessed of a sense that automatically registers the directional information and instantly computes the angles and distances, much as we recognize a color without having to indentify it by its wavelength or its position in the spectrum each time we see it; it has a *quality* of, say, greenness or blueness that we see and register. Likewise, its position relative to the sun has a *recognizable quality* to the insect.

How does the insect find its way on a cloudy day? Obviously, a complex sense of this sort could not exist without backup mechanisms. For the bee (to simplify our task by taking a single example) one of these is the nature of its eyes. Our eyes with their single lenses perceive and register objects by light waves that vibrate in all directions as they emanate from the sun. The bee's multifaceted eye, on the other hand, with its 6,300 lenses, perceives objects by polarized light, so that the picture registered in its brain is a synthesis processed from all these images, or partial images, that simultaneously convey the directional angles the bee needs.

What the precise sensory experience of the bee actually is, we have no way of telling. Polarized light comes into being when sunlight is diffused by the earth's atmosphere and then vibrates on only one plane. Because of this, some of the lenses of the bee's eye will see a full image, some a partial image, and some will not see the object at all.

The lenses of the facets (or ommatidia) of the bee's eye do not themselves have a polarizing capacity. There are deep structures behind each lens stacked side by side, but in each ommatidium at a different angle. Depending upon these angles, the polarized light will either reach the optic cell at the base of it or be partially or wholly blocked.

The bee's brain then makes its calculations, so to speak, by coordinating these images, producing information that is useful to the bee; that is, not only an image of the object according to the bee's needs (the image is not a photographic one, as ours is, but comprises essential elements of the view), but simultaneously the angle of the sun to the bee's flight direction.

The bee acquires a knowledge of objects in its environment on clear days and retains a memory of them. This memory serves the bee like a mariner's chart, so that on an overcast day it needs only a small patch of blue sky, one that provides a visual angle of no more than 10 to 15 degrees. The small amount of polarized light from that patch is then sufficient to give the bee the coordinates of its built-in mariner's map.

To appreciate fully the magnitude of the accomplishment of the bee or the ant by our standards, we have to remember that with every subsequent twist or turn in its explorations, with every new direction it takes, it automatically registers the new angle in relation to its previously irregular course. Indeed, for every new direction taken, its brain must calculate the product of the angle of the sun times the duration of that particular leg of the run, and then it must add all these sums and divide the result by the total running time. In this way it can arrive at an average movement angle to the light.

Thus when the bee's excursion is at an end, all it has to do is to turn around and follow the course of that mean angle in reverse. In effect it is as though the bee's brain were to contain a magnetic tape that constantly registers environmental information and can play back the information it has assembled to a calculation machine that digests the data and produces instantaneous solutions.

Miraculous as this seems to us, it is not all. Besides the bee's innate calculating ability, it also performs feats of memory well-nigh incredible for such a small creature. It memorizes tasks and performs them in correct sequence. If it is exposed to an experience that involves color (like a drop of sugar water on a colored dish) as little as three times consecutively, the bee will remember this for up to two weeks. If it has located a particular food site, this will remain in its memory for six to eight days.

Even more fantastic, it has the ability to memorize the celestial path and angular velocity of the sun. We know this because when the bee returns to its hive and communicates the whereabouts of the food supply to its hivemates by its waggle dance, it sometimes keeps up this communicative movement for two to three hours so that all the bees returning to the hive within that period may be informed. In the later stages of this performance the bee giving the news adjusts the angle of its dance movements to compensate for the change in the position of the sun that has taken place in the meantime.

The learning capacity of bees is impressive. They are able to learn signals in every known sensory modality, not only in those sensory modalities that are part of our repertoire but also in their own, which exceed ours. They can learn quickly and they can master multiple tasks that depend upon cues they receive from more than one sense at a time. Some bees have been trained to walk through relatively complex mazes taking as many as five turns in sequence in response to such clues as the distance between two spots, the color of a marker, and the angle of a turn in the maze. (Ants, by the way, can perform comparable feats.)

The bee's eyesight is also different in many other ways from ours. Because of the shape of its eye and its multiple lenses, the bee, like the fly, can see all around its body without turning its head. Therefore, unlike a human, it does not have a blind spot, and one cannot creep up and surprise it from behind. This kind of eye, of course, is marvelously suited to the necessities of the bee's life.

Although in some ways the bee's eye is more accurate than ours, in others it is less so. By our standards the bee is near-sighted. It does not see sharp images. Even large nearby objects look slightly fuzzy to it in about the same way that a myopic person would see them without his glasses.

A bee's sight also differs from ours in the selection of the images it registers. If we were to see through their eyes, even given no change in the rest of our senses, the world would be a completely unrecognizable place to us. It would be a bizarre world in which shapes would not exist as we appraise them, but would appear in broken patterns. We should not be able to distinguish between a filled-in circle and a filled-in square or triangle. All of these would appear the same.

On the other hand, if the shapes were interspersed with empty areas, like a cross, an empty square, parallel lines, or a Y-fork shape, we should be able to distinguish these from the filled-in areas but, again, not from each other. What catches the bee's attention, apparently, is the number of borders rather than the total areas of the various shapes, and this is understandable because sight of this sort is admirably adapted to finding a food supply secreted among the petals, stamens, and pistils of flowers.

The bee also sees a different part of the spectrum from the part visible to us. Its range goes beyond ours at the violet end but it does not see so far as we do toward the red end. When we look at a crimson flower, we see it in a different color from the bee, whose eyes pick up the blue-to-purple tones in its petals, whereas we see more of the red-to-yellow ones. Moreover, many flowers have markings along the inside of the petals that have evolved to guide insects toward the nectar. Some of these lines are in colors that we can see, but many are in ultraviolet colors that the bee can perceive but we cannot.

Karl von Frisch described the color world of the bees in these terms: a world with no red; where daisies, which look white to us, are bluish-green; where white roses, apple blossoms, bluebells, and daffodils glow in quite different colors. Some petals, he wrote, owe their lovely colors to an absence of ultraviolet light; in others, the addition of it is the source of a color magic hidden from us.

For instance, for us the yellow blooms of treacle mustard, rape, and charlock are hardly distinguishable in color and shape, but the bees know better. To them only the treacle mustard is yellow. The rape blossoms reflect a little ultraviolet light as well, which gives them a slightly purple tint. The petals of charlock reflect a lot of ultraviolet, so they look a deep crimson to the eye of a bee. Anyone who could look at the world through a bee's eye would be surprised to discover more than twice as many kinds of bloom as our ultraviolet-blind eyes can see, and with ornaments never registered before.

One of the possibilities that emerge from differences in the perception or nonperception of certain parts of the spectrum by the eyes of different creatures is a very sophisticated camouflage effect. To our eyes, for instance, both the male and the female of the Indian luna moth are light green and they are indistinguishable from each other. To their enemies they are hardly visible against the green leaves upon which they settle. But the luna moths themselves perceive ultraviolet light, and for them the female looks fair while the male looks dark. Nor is the green a camouflage for them, since they see each other as brilliantly colored against the, to them, grayish-green of the leaves.

Among the other ways in which the bees see the world differently from the way we do is one that is based upon the flicker-fusion rate. This is the number of flickers per second at which sequential images are no longer seen as separate. Human beings can distinguish from 16 to 24 flickers per second. Films that are to be viewed as motion pictures show about 30 "still" frames per second. A bee, which can see 265 separate flickers per second before fusion takes place, would see our ordinary movies as we would see a slide lecture—a series of still photographs. Thus the bee can see objects that are moving at far higher speeds than we can before the image becomes a blur or eventually invisible.

As if all this were not enough, in addition to its two eyes the bee also has three ocelli, eyelike structures that function like a camera's light meter. They record not an image but rather the intensity of the light. Of course, we human beings are also aware of light intensity in a somewhat crude way, but since we do not have organs specially evolved to bring constant and precise information of it to us, we have been obliged to invent the light meter for use whenever we need to measure light intensity exactly.

In giving this information about the special senses and sense organs of other creatures, we can, naturally, do this only in our own terms. We have somehow always to jerk our minds back to the thought that when we talk about "seeing" and "sight," we have a mental concept that is certainly quite different from what a creature endowed with a mind on a par with our own but serviced by senses like the bee's would conceive, if the subject of sight were discussed in its own language.

Of the other senses, smell and taste in the bee seem to be

approximately equivalent to our own, although some substances that are sweet to us appear to be distasteful to them.
(We must always bear in mind, of course, that we are speaking
of a creature that is about the size of one of our fingernails and
possessed of a differently organized nervous system.)

As regards hearing, the bee has no ears and is deaf to
airborne sounds, but it is moderately sensitive to groundborne
sounds that come to it as vibrations that are registered by an
organ in the lower part of its legs. It responds to vibrations of
from 200 to 6,000 cycles per second, which comes within the
low and middle range of sounds that are perceptible to us.

A bee's capacity for judging the texture of surfaces is infe-
rior to a human being's. On the other hand, it does have a
limited capacity to respond to the earth's magnetic field, which
we do not. Incidentally, our scientists have so far been unable to
detect any special sensory organ that mediates this sense in any
of the animals that possess it, including many of the migrating
birds and fishes and some amphibians as well as these insects.

An extremely well-developed temperature sense, also a part
of the bee's equipment, allows it to detect changes of as little as
one-quarter of a degree centigrade. Because of this acute sense,
bees are able to, and do, air-condition their hives by fanning
with their wings, maintaining a constant "indoor" temperature.

Moreover, the bees have an ability to sense changes in the
concentration of carbon dioxide in the air—a capacity that is
totally alien to us—and they also gauge changes in its humidity.
This last awareness is one that is also available to us, but in a
crude way when compared with the refinement of this sense in
the bee.

Bees have an excellent sense of balance; in fact, they have
gravity-perception organs that also monitor acceleration during flight. These organs, in their way, are as complex calculating mechanisms as one could imagine. They consist of hair
plates on the neck and legs through which flexible bristles
called sensilla grow, directly connected at their bases with the
nerve net.

As the bee moves, the bending of the limbs causes variable

pressures on the sensilla. This information is transmitted to the nerve net, which in turn conveys the composite pattern of the pressure to the information reservoir of the brain, which then automatically directs the necessary adjustments. The instrument panel of a jumbo jet airliner does not have more sophisticated apparatus than the tiny body of the bee!

A naive observer, told or actually observing all the wonderful capacities of the bee, might wonder why we humans need the comparatively huge size of our own bodies, and might even feel a sense of inferiority when comparing our own clumsy sensory skills with the finesse of the bee's. But it goes without saying that there are other distinct areas where the responses of higher animals are superior.

A bee can learn only skills that fall within the scope of its adaptation to its environment. Although these skills appear marvelous to us, the bee could not get along without them. Unable to calculate the angle of its flight to the sun on its outward journey while simultaneously accounting for the motion of the sun through the sky, it would get lost, and its society would fall apart. It has an absolute need for each one of its senses, whereas some mammals, if they lose the use of one or another of their senses, still manage to survive, if with difficulty.

A bee cannot be trained to a skill that is not native to it, but higher mammals can be trained to all kinds of alien accomplishments—as witness the circus's dancing bears, balloon-balancing seals, bicycling chimpanzees, somersaulting dogs, and porpoises that retrieve ladies' handbags from the depths of their oceanaria. It would seem that every last adaptive capacity of the bee has been called into play, while mammals still have considerable reserve capacities.

What is more, bees (or similarly endowed creatures) cannot reorganize the memories of their experiences to construct new responses in the face of a novel situation. In our personal experiments we have sometimes seen ant behavior that has looked like the epitome of human intelligence. Once, when observing the trails of ants to and from the entrances of their burrows on a clifftop habitat, we placed blades of grass and small twigs across or into the openings of their tunneled passageways, blocking the ants' entrance and egress.

We saw the first-arriving ants attempt to move these impediments and then, unable to do so, go away to summon help. Within a short time many dozens of the little creatures were at work trying to shift these obstacles, unsuccessfully. Then, after not too long an interval, to our great surprise, we saw our blades of grass being pulled down into the passages until they disappeared from our sight. It looked as though, having discovered that they could not move them by pushing or carrying them away, they had decided to get a better leverage by pulling them down into their nests and disposing of them there.

Anything that outwardly looked more like insight behavior would be hard to find, and we had to remind ourselves that the ants were merely using those resources available to them: pushing, pulling, and a drive to keep their passageways clear. The pushers had failed and the pullers succeeded—but the result was a clear passage and with all our brainpower we ourselves could not have done a better job.

On the other hand yet again, we have not only to see the sensory abilities of the individual bee or ant, wasp, or termite, but to view them in the context of the social group of which each is a part. Each of these tiny creatures has a brain commensurate with the size of its body, and a certain number of behavioral responses are genetically programmed into its nervous organization—probably as many as the available neurons can possibly carry.

But the sense organs it possesses and the marvelously adaptive responses initiated by its nervous system are exquisitely attuned not merely to the needs of the individual creature, but more importantly, to the needs of its social system as a whole. Several investigators have suggested that an individual social insect, whether bee, ant, wasp, or termite, is in effect not an entity of itself but rather a cell of a larger entity—the colony, hive, nest, hill—since its mode of existence and the role it performs is subordinated to the requirements of the entire group and indeed actually determined by it, in a manner strongly analogous to the subunits, or organs, of a total organism.

It is as though the cooperative presence of a certain number of these tiny creatures produces a mass that has a directing intelligence of its own—a quality that is present in the whole but not in any one creature singly, nor in the sum of the attributes of all of them as individuals. That brand-new effect, an attribute of their collective number, could not be predicted from the sum of the attributes of each of them.

With all this in mind, let us take ourselves on a journey to a planet where the perception of polarized light forms the basis of intelligent life, along with some special attributes of memory. This perception is not at all rare on earth. It is by no means unlikely that it may be found in other parts of the universe.

FIVE

The
Apistarians

ONCE AGAIN WE PROJECT ourselves into the future. This time we are on our way to what we have come to call the Green Planet, the world of the Apistarians. The overall green effect the planet gives from a distance is due to its particularly lush vegetation, which grows in such verdant profusion because of the almost perpetual daylight the planet enjoys. It orbits around a twin sun system, so that usually as its mother sun sets, it receives light from the more distant second sun. The "second sunlight" is not so brilliant as the home sun's, but the supplementary light is sufficient to foster the ebullient growth of all the plant life on this planet.

We arrive in an area that, but for the flattening force of our retrorockets, might have proved a handicap to landing. From far above, it looked like a green ocean, but as we descended through the atmosphere we discovered it to be a dense forest of giant, willowy, grasslike tufts growing to heights of 50 to 60 feet, so that our landing capsule was entirely engulfed in it after touching down.

As we emerged from our craft we immediately became aware of what seemed to be a sound signal being beamed toward us. We could not understand its meaning, nor did we know whether it was meant for us, but we decided to be guided by it and to follow its direction to its source.

We did not have much choice. The dense vegetation was pathless, and unless we took some arbitrary guide we should

71

have risked circling endlessly within it. Below the giant grass-trees grew a host of different kinds of plants of unimaginable variety bearing leaves, flowers, and fruits such as we had never seen before, as well as some that had bell-like, star-shaped, cup, umbel, or trumpet forms reminiscent of the flowers of our own world.

Among this wealth of vegetation we were impressed, too, with the great richness of insect life, some types growing to large, almost birdlike size. Here and there, as well, we saw small furry animals that bounded up and down the giant growth and leaped from plant to plant; we began to debate among ourselves whether they more resembled squirrels or tiny monkeys.

A little later, when we had an opportunity to see one of the "monkey-squirrels" more closely as it plucked a fruit and sat upright on its haunches to eat it, we could see why it could climb and leap, grasp and scamper, with such agility. All four of its paws had well-articulated fingers with leathery pads and strong, curved nails, so that they could be used as hands or feet indifferently: they could grasp branches, carry food, or flatten out for running.

In addition, the little animal sported a tail as long as the rest of its body, which it used to aid it in grasping and in balancing when some of its hands were otherwise occupied. To identify it to ourselves, we called it a monksell after the animals of our own world that it called to our minds. One of our party, feeling venturesome, plucked a fruit from the same vine that the monksell had raided, and ate it. He found it delicious, of subtle flavor, sweet yet tart, like a passion fruit. We all refreshed ourselves from the vine.

Fortunately, the abundant undergrowth was neither thorny nor too woody, and so we managed to make our way through it without undue difficulty. Shortly, we came across a small clearing, in the center of which stood a kind of stockade. From this point, three unmistakably hostile creatures emerged, making menacing gestures toward us. They appeared to be armed, although we could not immediately identify what seemed to be the weapons they were carrying. Later we learned that their lances were tipped and their propulsion guns loaded with deadly poisons derived from their vast knowledge of plant substances, and that flamethrowers and harnessed electricity were also part of their arsenals. But at that moment our only interest was to make it unambiguously clear that we came in friendship.

Our first reflex was to raise our empty hands to show that we were unarmed; then we held out toward them in cupped hands some of the fruit we had gathered. They seemed to understand our gesture as one that was intended to show nonhostility, because they allowed us to approach them.

Eventually we learned that this was an Apistarian sentinel group, one of a well-organized troop that guarded all the approaches to their settlements. The sound signal we had heard was rather a warning than an invitation. Apistarians had heard the arrival of our landing capsule and had gone to their observatory tower to investigate the source of the strange and alarming sounds. At first they had thought it was a meteor, but then when they saw the parachutes open and subsequently the billowing smoke from the retrorockets as they burned out the area for landing, they began to believe that perhaps hostile Apistarian forces had invented an incredible new apparatus for war.

After the initial excitement and discussion, they had chosen to remain calm but on their guard. In any case, they decided to set off the territorial warning they normally sounded when approached by strangers from other parts of their own planet.

One of the sentinels guided us to a locally responsible group of Apistarians. We were struck by a vague similarity between them and the little creatures we had called monksells. Not that they were really alike. In the first place, they were almost our own size—those we first met averaged about five feet in height—but something about the slightly rodent-like cast of their features and bodily anatomy was reminiscent of the forest creatures' in about the same degree that our own structure differs from and yet resembles the apes'.

We were also immediately struck by their eyes, which were not like our own but were ommatidian eyes with thousands of little lenses, each catching the light at different angles so that they gleamed in a way that was quite startling to us at first. It was like looking at the eye of a fly magnified a hundred times for illustration in a biology textbook, each lens a separate bead in the clustered whole. These eyes stood out more roundly from the face than ours do, and in proportion were about twice as large.

Most of the species on this planet had this type of eye, although we had not noticed it in the shade of the forest when we first saw the monksells. Later we discovered that our initial observation had been accurate, since the same relationship existed between the Apistarians and the monksells as between man and ape, in that both had descended from a common evolutionary ancestral species.

Among ourselves we debated the reason for their retention of a tail. After all, man and ape both have arboreal ancestry, but this appendage receded as long ages of life on the ground rendered it relatively useless. But as we came to know the life of the Apistarians better, we could see that the extreme lushness of their vegetation remained a factor in their adaptation to it, even long after their emergence as intelligent, toolmaking warriors living in culturally based social groups.

Any land they cleared had to be tended with the utmost diligence. After but a short period of neglect, their giant grasses took over their plazas and buildings with jungle-like rapidity. Thus, they always retained their familiarity and ease with the grass-trees of their origin. It was often simpler for the Apistarians to climb them and negotiate their high canopy with their own limbs, rather than constantly to be building many high towers and platforms that would have taken far too much of their time and energy to maintain. And so their lives were never totally divorced from their grass-trees as ours were from the forests in which our progenitors took shape.

At the time we visited them, the Apistarians were not an unsophisticated race. The most advanced mathematics came easily to them and they were both physically and mentally vigorous and adventurous. They had partially explored and were still in the process of exploring every nook and cranny of their world. In spite of the thick canopy of foliage that all but blotted out their view of the sky, they had climbed above it, eventually built high platforms (of the kind to which we had been taken to meet their authorities on arrival) in a number of places, and were familiar with the celestial bodies within their range.

Several of their leading minds were engaged with problems of flight. With this background it was not too difficult for them to accept the possibility of space flight, and as soon as communication was established between us they showed themselves insatiably curious about our techniques.

Before continuing we must pause to mention something about their type of sight and the special qualities of their brains that resulted from it. With the thousands of separate lenses that formed their eyes, their actual vision could only be compared in our terms with something like an exponentially increased polyphoto image of anything they see. Their brain, receiving this multiple image, computes it into a composite impression, so that what they actually took in at a glance was a considerably fuller picture of any object than we have.

Moreover, since direct lines of sight were hampered by the density and comparative uniformity of the vegetation in which they evolved, and since their ability to find their way was complicated by it, natural selection had preserved species on their planet that were sensitive to polarized light rather than to the light we see, which oscillates in all directions. Their most highly developed and intelligent creatures were among those whose sensitivity in this special area was most acute.

This sense, in addition to the constant pictorial computing imposed by their type of eyes, fostered the mathematical agility of their brains. Their sense of direction was absolute. Many circumstances that would have presented mathematical problems to us were not problems at all for them, because their eyes and type of vision demanded a brain function that interpreted angles, averages, and probabilities automatically and instantly. The result of this brain function could then be seen by them as a quality—as something that *is*, and not as a problem that has to be calculated.

In a society composed of creatures with minds to which the probabilities of success or failure in any undertaking are immediately apparent, the sensible members of this society will not be inclined to take slim chances. A fatalistic attitude will be engendered. The immediate perception of the precise degree to which a situation permits possible outcomes makes it less likely that individuals will devote energies to "the impossible dream" on a remote chance of realizing it.

We discovered that the Apistarian society, structured by such an awareness, had had to foster and make a special place for occasional deviants who would be prepared to persist with an endeavor in the face of a sure knowledge that it had little likelihood of achievement; otherwise the society would have become too static and no progress would have been made.

The workings of the probabilistic minds of the Apistarians was best illustrated by the type of warfare they waged. Much time and thought was devoted to strategy. It was concerned with maneuvering, with outpositioning the enemy, and as a result very little bloodletting ensued, because when one side became aware that it had been outmaneuvered, it gave up. It was like a game of chess with live chessmen whose movements were directed by their leaders and where, on the recognition of a checkmate, the loser concedes the game. Unlike the game of chess, though, the Apistarians waged their contests, not on our type of *logical* moves, but on their awareness of the probabilities.

In a way it reflected something of the dominance contests that occur all the time among so many of our own earth's species. Every device, bodily and behavioral, of which a creature is capable is devoted to display, to threat, and to challenge, but when faced with an opponent it *feels* to be superior, it submits without a fight and adopts subordinate behavior.

With the Apistarians, who were of high intelligence, the

results were not always quite so clear-cut. Sometimes one of them, perhaps able to coordinate more variables in his awareness of probabilities, would feel sure of a chance of which others were unaware. Such a situation did not occur often, but when it did, it would cause some confusion. Opposing troops would engage in all kinds of threatening display over prolonged periods of time, and for these circumstances the Apistarians had a special caste of umpires who would eventually be called in to judge a battle on fine points.

Incidentally, the Apistarians did not fight to enlarge their territory, but rather for mineral rights outside their own boundaries. Customarily they did not take prisoners when they won a battle, but exchanged their victory for mineral rights in territory controlled by the defeated army.

Their umpires were selected by nomination from among the leaders of each of the settlements. Every group sent two of its most esteemed "generals" to participate permanently on the "committee of umpires," and they were housed together on a certain island in one of their great bodies of water.

This brings us to the geography of the Green Planet. Although it was for the most part covered with great grass-tree forests of the kind in which we landed, the planet was also marked with mountains, oceans, rivers, and lakes, and, above all, many caves that spurred the curiosity and the exploratory propensities of its inhabitants. Their special orienting abilities made it inevitable that they should be explorers, and therefore, that they quite frequently came into conflict with other groups.

Minerals were very precious to them. Being the kind of creatures they were, anatomically totally adapted to above-ground living, it was only late in their cultural development that they began to recognize, first, mountaintops above the vegetation, then cliff faces, and finally caves as sources of minerals and ores. Soon these areas became the objects of warfare between communities.

Besides this, the planet had a few volcanos, which were the most hotly contested of all because of the sulfur, phosphorus, selenium, and other minerals they spewed out. The Apistarians also found and used an obsidian-like material and achieved marvels of domestic architecture using this for their main tools, very much as the brilliant civilizations of Central America did early in the current era on earth.

We were amazed, as we got to know it, at the wonderfully advanced technology they had achieved with the use of comparatively few basic materials. At the time we met them, their technological enterprise had been based mostly on the use of cellulose materials, but they had also used alluvial sands to fabricate glassy substances. The metals they had found at later stages in their cultural progress had been used, for the most part, ornamentally in the places where they lived.

It strikes us how late in our description of the lives of the Apistarians we come to talking about their buildings. In most places this is the first impression one gets of any population —indeed, one forms an impression of a village, town, or city from its buildings almost before one begins to form an impression of the people who live in them. With the Apistarians this was quite different.

Their buildings were set in the midst of so much foliage that one was almost on top of them before realizing that they were there. And yet, once in view of them, and especially inside them, we could not imagine how it had happened that it had taken so long before we noticed them. They were marvels of construction.

Two images kept recurring to our minds when we tried to relate them to anything we knew on earth: the first was the wonderful geometric structure of the beehive, and the other, the equally geometric but also artistically sophisticated buildings of the Mayas, the Zapotecs, and the Aztecs. It was immediately apparent that the Apistarians not only had mathematical abilities of a high degree, but were also workers superbly skilled in many crafts.

Their rooms of many geometric shapes were fitted together with arresting ingenuity, presenting constant variation and delight to the eye. Groups of these chambers were connected with other groups by complex passageways and galleries that wound between and around their grass-trees, sometimes incorporating these into their architectural design. The basic cellulose material had been processed in a great variety of ways: compressed to form rock-hard substances where needed; molded for decorative arches adorning their passageways; for bowers; for baths; for furniture of elaborate design that at first we thought to be abstract but later discovered was based on the forms of the flowers and fruits of their surroundings; or converted into filaments and woven into gauzy fabrics.

Glass was used almost playfully in these buildings, not simply to admit light, but also to catch and reflect it, to create lovely prismatic color effects, to glitter, and to bring the sight of the forest inside in many, to us, original ways—through floors or roofs, in friezes or corners—so that it became part of the interior decoration of the building. The metals and some of the other minerals they found or captured were also used in decorating their chambers; in these the advanced artistic skills of their workers were displayed, at their heights by far outdistancing our own great artists in this field, even a Cellini.

In their benign climate one would have thought that they would need little if any clothing, but we found that all of them, even their young, wore sturdy protective clothing at all times. The fabric of which it was made was subject to continuous research and improvement; it was all flame- and poison-resistant, whether it was produced in lighter textures for festive occasions or in thick, hard qualities to be used as armor in warfare. For festive use it was dyed in gay colorings with plant dyes, but as armor it was uniformly the predominant green of the forest for camouflage effect.

They had coal deposits but, interestingly, used the coal chiefly as a material for written communication, as we use lead or chalk. They also used mineral and plant dyes as we do, but coal formed the base of their everyday "pencil" and "ink."

They did, of course, need heat for their manufacturing, but although they were acquainted with the properties of electricity and were able to harness it, they obtained heat from solar furnaces. Using their advanced knowledge of plasma physics, they were able to produce and control extraordinarily high temperatures in their furnaces by this means.

With all this knowledge at their disposal, they had, none-theless, not discovered atomic fusion. Many of the discoveries they had made seemed to have remained laboratory curiosities with some practical application, but not fully utilized. We had the recurrent impression that they barely tapped their potential; this may have been due to the conventionalization of their social practices and the psychology that underlay and perpetuated these conventions in their day-to-day life.

Perhaps it also lay to some extent in their long-delayed maturity (which we shall discuss at greater length presently). Having spent half a lifetime in play, there was inevitably a period when they first reached maturity when their excellent minds were still subject to mercurial moods and to a lack of persistence. They usually had reached quite a high age in terms of their life span before the potentials of their often brilliant thinking were fully realized.

Then again, perhaps this condition was but a phase in their cultural development. We could not help remembering that on earth a Leonardo was occupied with the principles of flight in Renaissance times, several hundred years before the practicality of some of his ideas was realized.

In any case, the technology of the Apistarians as we saw it was chiefly concentrated on building, the chemical products of their plant and mineral resources, and astronomy. Astronomical observation and calculation were a never-ending source of challenge and occupation for them. Their view of the sky was quite different from ours, since their two suns and almost permanent daylight obliterated a view of any stars but the brightest and nearest, of which they sometimes caught glimpses when their second sun was most distant.

Also because of their twin sun system, the course of their planet was very erratic. It never duplicated its orbit in any individual's lifetime, so that exact computations of relative planetary and solar positions became their first and most absorbing science.

The Apistarians' own eyes gave them their original clues and prototypes for their astronomical lenses. They used not only alluvial sand for glass ones, but also quartzes for observatory instruments. With these materials they produced lenses of high quality, giving great degrees of magnification. Their solar furnaces were elaborations of the simple solar cell, and gave the intense heat they used primarily for their chemical production.

Spurred by their adventurous and warlike natures, they had initially turned to chemistry for their weaponry, developing poisons, tranquilizers, and nerve gases as sophisticated, precise in their action, and as varied as any to be found in the natural equipment of a large part of the animal world on our planet. But they had gone much further than do our animals, which use these substances through their innate processes to provide food for themselves and their young, and for self-protection.

The Apistarians consciously manufactured them, and produced them in quantities and qualities so lethal as almost to preclude their use. They became a part of their arsenal of threat, so dreadful that no one could bear to think of their being used. Their very deadliness contributed its quota to the bloodlessness and formality of their warfare. Nevertheless, the possibility that someday some group might indeed make use of these dreadful weapons was a sword of Damocles that spurred them to constant alertness in defense and constant endeavor to keep their own arsenals on a par with or ahead of all the others.

Had it not been for this threat of chemical warfare, like a dimly perceived but ever-present dark cloud on their horizon, the lives of the Apistarians would have been idyllic. Their high intelligence was not otherwise devoted to technological advance. Their environment was bountiful and they found no need to search for material goods to sustain life.

In this way again, they reminded us of the Mayas, who lived on the bounty of their land and who devoted the lives of their most intelligent individuals, and the arts and skills of them all, to astronomical studies and to ritual splendor.

For that matter, on our planet, the whales, too, have evolved huge brains, far greater than man's and with more complex structure, more convolutions, more neuronal synapses, and yet they have not used them for the accumulation of material goods that they do not need, but for social relationships and for communication. The same is true of the porpoises, whose brains are equal to man's both in size and in neocortical development. Of course, the anatomical structure of the whales and porpoises, especially their lack of prehensile limbs, has been a large factor in this turn of events, but in them we see that it is not brainpower alone that gives rise to technologies and civilizations of our kind. Hands are an at least equally important factor. A handicap to be overcome or a challenge to existence is another.

At any rate, the Apistarians were predominantly agile, lively, vivacious, and hedonistic beings. With no need to accumulate food for subsistence, their hoarding instincts found their way into pastimes. Private collections of anything and everything were to be seen in almost every chamber of their abodes: colorful minerals cunningly worked; leaf and flower specimens lacquered and preserved; precious stones, which they did not wear as jewelry but carved into exquisite small sculptures; examples of metalworking crafts; wood carvings; miniaturized indoor gardens, some very imaginative, like maps of their terrain made with dwarfed live plants; animal models and skeletons; utilitarian collections of plant extracts for ritual as well as for warfare; all kinds of fabric samples; ornamental clay and pottery utensils; and weaponry.

Most groups of Apistarians lived sufficiently near some body of water so that, besides the small animals of the forest and some succulent types of grubs, fish were also available when they wished to vary their chiefly vegetarian diet. In spite of their generally plentiful sources, however, they took the precaution of keeping reserve stores, largely because of the irregular seasons, which meant that there were sometimes longer intervals between periods when certain foods were available. Even their days were not of regular duration, but depended on the current position of the planet in relation to the second sun. This circumstance affected all life on the planet fundamentally, not only in the length of their day, but also their tides, the magnetic field, and all the implications of these irregularities. A propensity to hoard was therefore not only selectively retained in their genetic endowment, but also culturally reinforced.

The Apistarians, traditionally valuing the accumulation of foodstuffs and necessary materials, came later on in their cultural history to place a high value on those collections that were originally a kind of decorative and playful extension of the habit, and then became objects of value of themselves, endowing their owners with social privilege according to the refinement and rarity of their collections. Thus, social standing among the Apistarians rested upon two bases: firstly, martial qualities—and this with them implied a particularly high standard of accomplishment in probabilistic mathematics—and secondly, on the quality and housing of their collections.

This duality of their social ideal in some way affected, or perhaps stemmed from, their idea of the world as dominated by the twin suns, which they celebrated as the sources of the bounty of their lives and to which they devoted colorful rites at certain times when the suns approached each other most closely. In our terms these rites had a religious quality, since the Apistarians identified their home sun as a symbol of the female element of life and the second sun as the male element. But for the Apistarians our concept of religion was alien. There was no mysticism in their ritual and no dogma. Their celebration was more of life than of a creator of life, and consisted more of social display than of worship or propitiation.

A duality was also apparent in their personalities. The great intelligence of the Apistarians was based on an inordinately prolonged youth. Their young were born in an extremely immature condition, almost like the aborted embryos of our marsupials. But the Apistarians had no bodily pouches to protect them. They reserved special rooms in their abodes connected with their living quarters by passageways but separately maintained, the temperature and humidity controlled, and with individuals constantly in attendance to care for the infants until they were fully developed and old enough to be able to move around in the grass-trees with the other young.

The Apistarians were not, strictly speaking, mammalians. They had small gland pouches on the inside of their mouths that produced a substance we could equate, perhaps, with pigeon's milk, and which they fed to these almost larval young. Both males and females had these glands and produced this substance, which is formed by the breakdown of cells, just as milk is in mammals. Later, as the young developed, they were fed mouth-to-mouth by either parent until they reached what we should consider as childhood and were able to fend for themselves.

Their childhood, too, was extremely prolonged. Sexual maturity had become progressively delayed in their species until it had come to coincide with social maturity. On earth in apes and in man about one-third of a lifetime is spent in the youthful phase, although sexual maturity occurs in both at a markedly earlier age, but in the Apistarians the youthful stage both physically and socially had come to last for half their lifetimes. We imagined that this development must have had to do with their rich food sources, which might otherwise have created a tendency toward overpopulation.

As it was, about half the population consisted of agile, active, frisky juveniles, at home in the high canopy of the great grass-trees, which they were able to clamber up and down and swing in with the greatest dexterity, coordinating the use of all four of their limbs and their tail, and exercising their optical judgment, their orienting abilities, and their muscular strength while doing so.

Yet shortly after these lively young became mature members of their groups, their preoccupations and way of life underwent a great change. Although they always retained an ease in climbing, they then spent virtually all their time on the ground. As adults they became aggressively competitive, trying to outdo each other in their hoarded supplies, in their collections, in the decoration of their abodes, in their explorations, in their efforts to be chosen to perform tasks that would give them access to the mineral rights that they all coveted as a source of supply for those objects that brought them social recognition and status. Which brings us to their economy.

Theirs was an economy in which the necessities for subsistence were all around them. For food they had but to gather the vegetable products or grubs as they required them, to make an excursion to the nearest river or lake for fish, net any small animal of those all around them, or to draw on their stores for out-of-season supplies. They had no need to exchange goods.

When we reached a point in our contact when we were able to communicate with them and tried to explain to them our money economy, it was the most difficult concept for them to grasp. They found it hard to believe that turning mineral wealth into small metal disks for exchange against foodstuffs, clothing, housing, and the other necessities of life could possibly work. "Why do you have to exchange metal for these things?" they asked us. "Why do you not grow your own trees, make your own clothes, build your own buildings, and keep the beautiful metals for your honor?" We attempted to explain the principles of the division of labor to them, but they only responded to the effect, "How boring it must be to spend all one's time doing one thing."

They had manufacturing enterprises, concerned chiefly with the making of weapons, instruments for their observatories, glass and compressed cellulose materials for their buildings, other cellulose materials for their clothing, and, above all, chemical products, but these were all community matters. The materials were distributed where they were needed, not obtained by exchange for other goods.

Workers, directors, and researchers were easily obtained, because to choose an individual for any of these tasks was the greatest mark of respect the community had to offer. The privilege of doing the work was the object of strenuous competition and social maneuvering, especially since all work for the community was rewarded by exclusive rights to explore certain areas for their mineral content, and the gems and ores extracted became the symbols of prestige.

One of their favorite hobbies also contributed to general well-being. Very many of the mature members of the groups devoted a great deal of time to plant genetics. When they were successful in producing an unusual variety, they permitted the other members of their own group to help themselves from it, and this, too, contributed to their social standing.

There was another area in which the Apistarians' view of the world was quite different from our own, and in this case we mean literally their *view* of it. We have already mentioned that Apistarian eyes were of an ommatidian type that registered multiple images which, when coordinated in the brain, gave them a total perception of an object. Later we came to learn that their perception of polarized light gave another dimension to their vision.

We first got a clue about this when we were trying to master some words of their language and found that they had dozens of different words for what to us was a single object—say, one of their grass-trees. Slowly it dawned on us that a time element was an integral part of their vision. They never actually "saw" a grass-tree in the same terms as we did; it had separate existences for them as though it were a different thing at different times, determined by the angles at which the light from their suns reached it.

What we saw as a particular grass-tree they saw variously as a one-o'clock grass-tree, a five-o'clock grass-tree, or a ten-o'clock one; the different names incorporated their perception of the time element. What it really amounted to was that for them time was fused with their perception of an object.

While their eyes actually saw objects in disparate bits and their brains coordinated these, simultaneously their brains also coordinated with the sight of an object their perception of the suns' positions. This was a particularly important part of their awareness of their environment because, having no regular cycles of days or years, they could only refer to the positions of their suns for a concept of passing time, distance, and direction.

As we mentioned, the Apistarians were a lively, active race. In youth they all walked, hopped, ran on four limbs or with the aid of their fifth (the prehensile tail), climbed, walked upright, or swung along by their arms. They could also swim. No method of locomotion was alien to them. In their mature years they usually walked upright in a dignified manner, but they always retained their ease in the other modes and occasionally used them. Yet in spite of all this bodily activity their active minds also had outlet.

When we first saw their writing we thought we should never in a lifetime be able to find any coherence in it. Their symbols consisted of disjointed lines in which we could find no consistent design. We could recognize neither particular nor overall patterns to it, and could not see how it related to their speech, either by ideographs or by symbols for sounds.

As we attempted to explain our problem to a group of their learned individuals, one of them, with considerable insight, found a solution. Glass lenses that screened out all but polarized light were made for us at the center where their observatory equipment was fabricated. The writing was then put on a rotating cylinder so that, viewed in movement and through the lenses that simulated their own type of vision, we were able to see the writing as they saw it naturally—and then we could see its forms.

Suddenly the small sculptures we had seen in their beautiful miniature gardens also made sense to us. We had seen them as disjointed edges and outlines before, and wondered at the aesthetics that had produced them. But when we returned to them and rotated them sharply, with the aid of our polarized light lenses we then saw them as the natural forms now familiar to us. It was in a way similar to those small charms that our girls sometimes wear, with the letters of a message broken up into two planes, but that, when we spin them, merge and we can see clearly. The difference was that the Apistarians' sight could take in several planes and read them as they were, since their brains automatically did the coordinating.

After that, with the help of a young Apistarian who seemed to enjoy guiding and explaining their ways to us, it did not take long to decipher their symbols. It did, however, take much longer before we were able to gain an idea of the concepts they were expressing, since with their kind of eyesight their view of the world was so different from ours that this could not but affect their ideas about it.

Besides the fact that every object had multiple names, depending upon the angle of light in which it was seen, their different thought processes were revealed in the structuring of their language. Thought, for instance, was movement in their terms—one of many kinds of movement, and expressed with words similar to their words for running, leaping, climbing, seeing, and knowing, all of which, to them, had an element of motion in them.

It seemed strange to us that we found very few words dealing with emotions or moods in their language, and quite a long period of observation passed before we discovered the reason for this. In addition to their spoken language they had a huge vocabulary of purely gestural communication, which expressed all their feelings and emotions or moods. It was, to us, vastly complex. Thousands of gestures utilized the face, the limbs, actions, all parts of the body, and complex modulations and combinations of these to convey the subtlest nuances of meaning. All their personal lives and much of their public life were regulated by this nonverbal expression.

The extreme importance of what we can only transliterate as "honor," or "social standing," called for a perfect knowledge and use of these thousands of gestures in the highly ritualized formalities of their adult lives, and a very large part of the long period of their youth was spent in mastering them. Thus their equivalent of books was less important to most young Apistarians than to our young. Classrooms had no place in their lives. They learned by observation and imitation, and when they came to be chosen for special tasks in their communities, they learned these by apprenticeship.

On the whole we found it much more difficult to relate to the Apistarians on personal terms than we had to the Olfaxes. The Olfaxes' keen sense of smell was of another order than ours, but it was a faculty of which we ourselves had some vestigial capacity and so it was not alien to us.

With the Apistarians it was different. They saw things we did not see, and the things we saw were not perceived by them. Their mathematical minds drew instant conclusions that we should have needed a battery of calculating machines and equipment to reach. Their technology was in some ways superior to our own and in other ways nonexistent by comparison. We came to understand the most salient of their ways of life and basic principles, but we should have needed a lifetime of study to have gained a real personal feeling for its finer points and the wholeness of it.

We had, almost equally, two contrary impressions of the Apistarians. One, influenced by their idyllic and providential environment, was that they might well remain as they were over long eons of time. After all, our own Pacific Islanders, in similarly benign surroundings, although they explored the oceans and lands within their reach, never evinced any desires to conquer the world.

The other impression was quite contrary. When we thought of their competitiveness, their constant alertness against possible attacks, their curiosity, adventurousness, their extraordinary technical abilities in areas like astronomy and chemistry, their remarkable gift for orientation, and their drive in their search for minerals, then we thought of them as beings on the verge of an expansion into new horizons. If we tried to make analogies in our own terms, we thought of the seaboard peoples of Western Europe in the sixteenth century, whose similar qualities and whose drive to search for spices began with the exploration of their own world and ended a few short centuries later on the moon—if, indeed, it can yet be said to have ended there.

SIX

The Plutonians
and the Hydronians

LIFE IMPLIES SELF-REPLICATION, and self-replication implies specific chemical reactions between chemical compounds.

Because the chemical compounds that make up living cells are not inert but do react in certain ways, the simplest one-celled living organism has an ability to propel itself (or, more accurately, of its nature propels itself) toward or to withdraw from certain stimuli like light or heat, according to whether these are congenial or uncongenial to its well-being. In other words, living material, of its nature, interacts with its environment and is responsive to it.

With this fundamental property it can be seen that *all* the potentials of life are incipient in the simplest organism and that, given suitable environments, some or others of those potentials will probably eventually become manifest.

This applies to all aspects of life, but our present concern is with the senses. The senses are, of course, the ultimate elaboration of this chemical interaction between environment and organism; which type of sense eventually becomes the predominant gatherer of information for a living thing will ultimately depend upon the type of environment in which, over generations, it happens to have been formed.

Our planet with its varied terrain—mountains, plains, oceans, deserts, rivers and lakes, icy wastes and fertile valleys, hot, humid tropics and barren, oxygen-poor uplands—provides as good a laboratory as can be found, certainly anywhere in the solar system, for the study of most, if not all, of life's potentialities.

We now invite you to join us in exercising our imaginations within the boundaries only of what we know to be possible for life from our own experience of it. So far we have dealt with senses familiar to us—smell and sight—although we have taken aspects of these senses that are not at our own service. Let us now take a look at a sense that is completely foreign to our own methods of perception.

Electricity is a force used by many of earth's creatures to obtain an impression of their environments, and it is entirely possible that this should also be the case in other worlds. When we speak of electricity we are not referring to man-made artifacts, but to the impressive, naturally occurring sense mechanisms that have evolved on earth, especially in creatures that live underwater.

That this should have happened is not really surprising. Electricity is part of the very stuff of life: the biological activity of every cell is either accompanied or promoted by electrical changes. Little wonder then that electricity has been harnessed to the service of the perceptions in many species. Indeed, it is inevitable that this should have happened, because any energy that can be interrupted or modulated in its intensity can lend itself to use as an information-gathering mechanism by a suitably constructed sense organ, and anything that *can* be used by life, sooner or later *will* be used.

Electricity can be used by organisms in a variety of ways. It can be used as an emission, sending out informatory signals. It can be used as a weapon to stun or kill prey or enemy. Finally, it can become part of a total sensory apparatus by means of which a creature perceives its environment.

The electric sense can take alternative forms. Like the echolocation of bats, which registers the reverberations of high-frequency sound waves, electric impulses may also be emitted to touch or surround objects in the vicinity and be echoed back to the sender. Alternatively, an animal may surround itself with an electric field so that any object intruding into this field distorts it, thereby becoming apparent to the creature at its center. The essential difference between these two types of electric sense is that the first is sent out when needed (the bat does not emit its high-frequency sound when it is at rest), but the second is maintained as a constant state—like a private fog within which any object creates a distortion that can be perceived.

In our attempt to construct a world that could logically be based on electric sense perception, there are three areas we have to determine before we start out. The first, so that we may have a base in reality, is to discover exactly how these senses are used on earth; the second is to define the kind of environment that will necessarily have led to the selection of electric sense perception above all others; the third is then to attempt to obtain an idea of how the information gained in this way is likely to be integrated in the mind of a being receiving it—of what, in fact, is its idea of the world it inhabits.

What do we find when we look for electric senses here on earth, and how do they operate?

On earth true electric organs, as opposed to biolumines-cence, have evolved only in fishes, but in them these organs have evolved independently in six different groups—which would indicate that there is a powerful tendency toward such electric organs emerging among the mechanisms of life. Some of the more familiar fishes that possess them are the electric eel, the electric catfish, the electric ray, the stargazer, the knife fish, the skate, and the elephant fish. Besides these there are many more that are less familiar to most people.

A zoologist named Henning Scheich, after an in-depth study of one fascinating species, the Eigenmannia, reported about its social behavior. When two of the creatures meet, they have the ability to adjust their frequencies up or down to avoid jamming each other's signals. If we were to communicate by humming instead of by discrete, differentiated sounds, we should have to make a similar adjustment if a person tried to communicate with us on the same pitch we ourselves were using; otherwise we could not hear each other.

To do this electrically is quite a feat. In a way it is like modulating one's voice to produce different intensities and levels of sound; for us this is rather easily accomplished by adjusting the intake and expulsion of air into and out of our vocal apparatus. The Eigenmannia, becoming aware of another signaling at close to its own frequency, must alter the frequency of its current, presumably by chemical means governed by nervous impulses.

Dr. Scheich showed the versatility of the responses of the Eigenmannia's electric organ by putting an electric dipole into the fish's tank and adjusting it to the frequency of the fish, whereupon the fish changed its own frequency either up or down. This test was effective over a very wide range—in its way, like a talented singer with a wide vocal range.

To respond in this way, the fish, like a fine piano tuner, must be able to detect how near the extraneous stimulus is to its own frequency and also whether it is higher or lower. It must do this by analyzing the coming in and out of phase of the two waves that show a slightly different frequency, since the number of times the two waves come into phase with one another is the difference between the two frequencies.

As a further refinement, the fish's discharges are not pure sinusoidal waves—on the contrary, the wave shapes are rich in harmonies, enhancing their possibilities as a system of communication. Scheich actually found particular neurons in the fish's midbrain that responded to these differences in electric frequencies, which are entirely analogous to the centers of language in our brains.

Another researcher, P. Moller, working with Mormoryd fish, found that these have a specialized lateral organ that emits an irregular discharge that becomes regular as soon as another fish approaches. It has three types of responses: one is complete cessation, when the fish is electrically undetectable; it may evoke this response either to hide or to "listen." It can also vary its frequency or its frequency regularity. When a Mormoryd fish intrudes into the territory of another, the home fish raises its frequency to assert its dominance—almost like one human being shouting down another in a verbal exchange.

T. H. Bullock described similar patterns of social interaction in the Hypopomus, which lives among dead leaves in the turbid forest streams draining into the Gatun Lake in the Panama Canal Zone. He noticed that the Hypopomus has a pacemaker in the medulla oblongata (where the brain is joined to the spinal cord) that operates somewhat like the pacemaker of the heart in man, regulating a private frequency for detecting objects while it uses another for communication and warning.

Quite recently, Dr. Carl Hopkins of Rockefeller University spent a season on the banks of the Mocomoco Creek in southwestern Guyana, "listening" for the discharges of the numerous species of nocturnal electric fishes that live there. He placed wires in the water, and if a fish swam near them, its electric discharge was picked up, amplified, converted to sound, and emitted through a transistor radio speaker so that Dr. Hopkins could, so to speak, hear what the fish was saying.

He noted that when a female Sternopygus swims past an adult male of her own species, she quite literally "turns him on," for suddenly his steady single-frequency discharge becomes a chaotic electric "love song." He also noticed that each of the closely related species living in the same waters is characterized by a different kind of electric discharge that helps individuals to recognize their own.

In the species Sternopygus macrurus he found that although immature males and females signaled at similar frequencies, the signal of the adult male was distinctly different from that of the mature female. When two mature males swam near each other, nothing much happened. They both continued to emit a steady discharge which, when translated into sound, "resembled a bassoon stuck on a low-register note." But when a female in breeding condition entered the electric field of a mature male, the male's steady drone changed abruptly to an electronic chant. In fact, when Dr. Hopkins imitated the female's signal with his equipment, the male serenaded the wires suspended in the water.

All these researchers, as well as several others, were following up observations originally reported by H. W. Lissman of Cambridge University, in 1958. Lissman kept two Mormoryds in a single tank but on opposite sides of a cloth barrier, so that they were denied visual or tactile information about each other. Nevertheless, they became aware of each other, and Lissman determined that it was by means of low-voltage emissions (about 1 or 2 volts—too small for us to detect without amplification) in which the fish established a current-density field resembling an electric dipole around each of them. He suggested that this might play a role not only in social behavior but also in navigation in darkness.

We would mention the electric eel only in passing, because it uses its electric organ merely as a weapon to stun or kill enemy or prey and not for communication. It possesses some 500,000 electric plates, modified muscles that discharge up to 550 volts and 2 amperes, which it can do up to 150 times an hour without visible fatigue. Indeed, it is so powerful that it has been known to kill a horse fording a river. The electric organs of the eel contain columns of large cells stacked one next to the other in order to attain this power; this construction is said to have inspired the concept of placing man-made electric cells in series of rows to produce a voltage that is the sum of the individual voltages of the cells.

In other species of electric fish, including, for instance, Gymnotus, Eigenmannia, or Apteronotus, however, the electric receptor organs are not modified muscles but rather organs that have evolved from nervous tissues modified from the lateral line organ, an organ of perception that has developed in most fishes for analyzing the environment by means of pressure. They enable the fish to register and interpret lines of force so finely that it can recognize a glass rod only 2 millimeters in diameter (almost invisible in the water), and distinguish between two objects of the same size and shape but made of different materials.

Having looked at the way the electric-field sense is used by some creatures on earth, we now come to the question: What kind of an environment would have to exist on another planet for electric rather than other senses to have evolved in a species of high intelligence?

Obviously, the first essential is that the place is dark. It is not enough for it to be cloudy, heavily overcast with a gaseous atmosphere, or partly darkened by long periods of night, because we know that eyesight can be modified to be useful in all these circumstances, especially when in combination with other keen senses like smell and hearing. Many animals on earth can see at night by the light of moon or stars; indeed, we ourselves can become quite adept at picking out familiar shapes at night, once our eyes become accustomed to the darkness and can discern shapes made familiar by our daylight view of them. We can also see objects on heavily overcast days even if not as precisely as on clear ones. This is not the answer.

On earth, creatures that do not use sight at all are those that live in total darkness all, or almost all, the time. Bats that live in the deep recesses of dark caves have no use for sight, and so they have evolved other, although not electric, senses; neither have certain fishes that live in caves or in murky, vegetation-filled rivers. By far the largest number of species that use electric senses are those that have their habitat either in the great depths of the oceans or in turbid, fast-flowing fresh water.

At this point we take note of a difference in the types of electric senses brought into use in these various habitats. The fishes that live in caves orient themselves, like most other fishes, by their lateral line sense, moderated by the organ that runs along their sides and registers the surrounding environment by sensitivity to its differential pressures. This sense is not an electric one, any more than is the bats' echolocation.

Deep-sea fishes, however, do use light signals to identify themselves to each other. These light signals are a form of bioluminescence, which is sometimes used by dry-land creatures also, like the Central American click beetles and the European glowworms. The organs that produce bioluminescence are a variant of those that emit invisible electric charges, like the electric-field organ we have just described.

The knife fish and elephant fish, however, are not able to use their lateral line organ effectively because the turbulence of the waters in which they live denies them a stable environment. The motions of the waters that surround them, registered by their lateral line organs, would disorient them, were they to rely solely on awareness of external pressures for information. Therefore the electric-field apparatus becomes essential to them.

This apparatus is extremely sensitive. Professor Lissman showed that it reacts to a drop in voltage as small as 0.03 millionths of a volt per centimeter. This sensitivity is so fine that one can only compare it with that of an eye that can perceive a single quantum of light or an ear that registers vibrations of subatomic magnitude. The sensitivity of the receptor organ is matched by the brain's interpretation of this electric-field data as it is received from all over the body.

By far the largest part of the brains of these fishes is concerned only with the processing of the electric sensory stimuli. In some the electric area has outgrown all the rest of the brain, much as the neocortex—the reasoning brain—has in man. It is quite interesting, in this regard, that elephant fish are the only ones that show a propensity to play, and since playfulness is associated with intelligence, it would seem that, given propitious circumstances, the elephant fish would be capable of developing, in the course of further evolution, an intelligence perhaps even as great as the dolphin's.

In conjuring up worlds where these electric senses would become the tools of a high intelligence, then, the first prerequisite is that they must be dark all the time. If the world is dark but with a still atmosphere, we are most likely to find forms of life that inform each other of their presence and kind by means of bioluminescent signals, but that receive other information by a pressure sense, an echolocation sense, or by smell and hearing. If we would conjure a world of electric-field perception, it must be not only dark but also turbulent.

Under what conditions would a world be dark at all times? There are but two alternatives. One is that it is entirely covered by a great depth of water; the other is that it is so far away from its sun that it is provided with no more light than perhaps a bright star like Venus or Sirius provides in our sky.

Very deep water seems to us at first to be not too likely an ambience for the development of highly intelligent life. The most intelligent sea creatures on earth are the marine mammals that have to remain near the surface in order to breathe, all of which have seeing eyes. Besides, aquatic life tends to streamline anatomy and usually to eliminate hands, which we have discovered are almost, and perhaps equally, as important as brains in the evolution of a species capable of culture and technology. True, the sea otter is playful and intelligent and uses its hands to gather food and to pick up and use tools, but its ocean habitat precludes the more complex aspects of technological development.

We are left, then, with a planet that must be the outermost of a solar system, orbiting at a distance from its sun so great that it receives no light from it, and where the atmosphere is so turbulent that other senses like olfaction or pressure-sensitivity would be disturbed. We do not yet know whether the surface of Pluto in our own solar system would support life, whether its surface is solid, whether it has an atmosphere that could sustain or whether its temperature is at all conducive to life, but if our real Pluto will not sustain the Plutonians we are calling into being, then they must inhabit the Pluto of another solar system.

The next requirement we have to supply is heat—not an excessive heat beyond the tolerance of organisms, but a temperature range in the moderate band. In the absence of sunlight this heat has to be engendered internally. If it were produced by atomic energy from the planet core, any eruption would be likely to wipe out life. Rather, it seems to us that volcanic action and its concomitant thermal springs and geysers would be a more likely source. Moreover, the gases emitted by volcanos might originally have provided a basis for their atmosphere.

We now come to a larger problem.

Most science fiction writers who try to map out other worlds confine themselves to technological possibilities. When they do speculate about other intelligent forms of life, they either assume these to be mechanical super-robots or living forms that are pure childhood fantasies. In any work of science fiction we have come across, we have never discovered the invention of an intelligent life form based on anything remotely possible, given life's processes. Even one of the best known of these writers, Arthur C. Clarke, in his *Profiles of the Future,* wrote:

Nowhere in space will we rest our eyes upon the familiar shapes of trees and plants, or any of the animals that share our world. Whatsoever life we meet will be as strange and alien as the nightmare creatures of the ocean abyss, or of the insect empire whose horrors are normally hidden from us by their microscopic scale.

Why "nightmare creatures"? The fantasies of dreams are connected with childhood's fear of the unknown and not with any possible reality. And why an "insect empire"? Indeed, such a world is hardly possible. High intelligence demands a central nervous system, while the insects have ganglionic nervous systems that are not capable of the integration and reasoning processes necessary for the functioning of high intelligence in individuals.

Then again, a highly intelligent form of life cannot have come into being unless lower life forms were also present—and so on down the line until we come back to plant life. In effect, the whole animal kingdom is a parasite upon the vegetable kingdom. It cannot exist without it. The animal requires the vegetable kingdom to synthesize organic from inorganic matter and thus to sustain it. Even more important, it was the respiration of vegetable life that on earth provided and sustained an atmosphere within which animal life was viable.

So now we are back to the atmosphere, and if we have an atmosphere containing oxygen, we almost surely must have at least some kinds of plants. This is vastly complicated by the absence of sunlight, so we turn our attention again to the ocean depths, to caves, and to the other dark places of earth's surface. In all of them we do find some kinds of plant life, and we must use these as paradigms for the vegetation of our Pluto.

Before we follow these thoughts through and begin to imagine and describe the physical world of our Plutonians, we must first fulfill our third requirement and attempt to visualize how a world, whatever its nature, might be perceived by an intelligent mind that receives its information via electric senses, and how that mind might integrate such information. Only then shall we be able to visualize what kind of being we might encounter and what form its social existence might take.

We have to keep in mind that no matter how exotic the mechanism by which a mind gathers information, its intelligence is based upon the manner in which the brain processes it. To a certain extent, whether we perceive an object visually, by touch, by hearing, by smell, or any other sense, the end result is an impression formed in the mind of the nature of that outside object. In many cases, the varied means may lead to a similar end result; but to a being endowed primarily with electric senses, the world must nevertheless appear fundamentally different in many ways from the way the same world would appear to us.

For example, electric currents have properties different from visible light, and so many things that for us are opaque may be transparent to such a being, and vice versa. Light waves, for instance, are reflected off the surfaces of objects, whereas electric currents may penetrate them to varying depths depending upon the nature of the material, particularly its conduction. It is quite possible that we might be perceived by the Plutonians approximately as we ourselves see an X-ray picture. But of course, if we were accustomed to seeing other people with X-ray eyes, our brains would certainly in time and with experience provide us with just as much information as we now gain with our type of vision. That is to say that just as we see the varying colors of light, the Plutonians might be aware of varying textures and other properties of materials by means of a subtle shading of electric conductivity that might have for them approximately the same value that awareness of color has for us. Such speculation would find some confirmation in the behavior of Lissman's Mormoryd fish, which were fully aware of each other in spite of the thick cloth separating them.

The appearance of the external form of things would not be the only difference between our type of awareness and theirs.

Deep differences would also exist between their and our awareness and expression of moods and feelings. Their perception and their communication would have to be governed by the electric waves that they emit and receive, and these would probably not be smooth, but would be formed of specific and characteristic, although minute, undulations modifying the overall wave pattern.

In terms of our own perceptions, it is as though one were to strike a pure note on a tuning fork and then repeat this note on a violin string. Although the note is the same on both instruments, we hear it differently because of the overtones produced by the violin string, and for that reason we can tell the tuning fork from the violin string aurally.

The awareness of our Plutonians, because of the wave nature of their primary sense modality, would therefore be closer to what we might think of as the flowing of a song than to the discrete images formed by our own verbally expressed thoughts, based on our visual perception. As a matter of fact, some investigators have expressed the belief that in our world the memory of homing pigeons is formed a little in this way—that as they fly over the features of a landscape, these features impress the pigeons like the themes of a song, forming an harmonious whole that sustains their recollection of them.

If we imagine a world populated by creatures of high intelligence, capable of advanced cultures and technology, whose sense perceptions have evolved along lines such as these, and if we are to imagine them when they are as far along their evolutionary path as we are on ours, we shall have to take it into account that their brains may be even more complex than ours, or in any case that they are capable of refinements of awareness beyond our powers.

We have to envisage creatures sensitive to all kinds of electric fields, which exist in abundance and almost everywhere here on earth, but from which we can draw very little information because we do not perceive them without the aid of instruments—and instruments are usually crude when measured against naturally evolved innate senses. Such beings might be able readily to perceive what we call "auras," and perhaps find quite normal a knowledge of things that we believe can be gained only by ESP.

Moreover, they might be able to pick up radio waves and thus receive, and perhaps also send, information over great distances without any need for instruments. A few strategically placed relay stations might extend this perception and communication in a way that is almost impossible for us to grasp. To get some feeling of what a sense like this implies, we must think of our heads as receiving and broadcasting stations sensitive to all the electric impulses all around us and at the same time able to respond to them and send our own messages.

To a certain extent our own brains do in fact do something like this, because they convert the messages received via our senses of sight, sound, taste, touch, and smell, into electric impulses that they interpret. But while our brains produce electric messages, they do not receive them directly except in the grossest forms, such as electric shock. Sometimes, it is true, we do become aware of the presence of an electric field when we happen to interrupt one, by means of a tingling sensation in the skin; but having no discriminating receptor organs, we cannot draw any information from it. Even when electric or similar wave emissions are used deliberately in the course of, say, medical treatment, we are still almost unaware of them, feeling at the most a certain degree of heat, but sometimes not even that.

We now have in our minds quite a lot of information about what the world of the Plutonians with their special electric senses could be like. It is dark; its atmosphere is turbulent; it is a great distance from its sun; its vegetation must have adapted to the absence of light in the way that our vegetation in the depths of caves or oceans has done. Its lower forms of animal life may use electric senses of a simpler nature and bioluminescence.

We know that for high intelligence to have evolved, useful prehensile organs as well as complex brains are necessary. We also know that the species developing this intelligence at some critical stage in its evolution probably had to overcome some inferiority in bodily strength or anatomical specialization vis-à-vis other animals inhabiting the area where it arose, or the forces of selection would not have isolated and spurred its incipient intelligence to the heights of overcompensation.

Let us now try to describe the world and imagine something of the culture that could have been developed by these beings.

To begin with, if we would wish to have any communication with them at all, we should have to start out on our journey to this world with a set of instruments that would duplicate their sense organs. The darkness and atmospheric turbulence in which life there must have developed would render useless any of our own distance senses and most of the instruments we use to extend them.

Radar, for example, would show the turbulence but not the individual creatures or objects within or beyond it. Our distant sight would be impeded by the darkness, and sound would be deflected by the eddying atmosphere. While we might consider using infrared light to see in the dark, the ever-shifting atmosphere would also interfere with useful distance reception by this means.

On the other hand, most of our proximity senses would be useful. Our olfactory sense would still serve us over short distances, but would not give us quite as much information as it normally does because we would have no way of interpreting it. The Plutonians would probably possess some form of vision adapted to darkness and to short distances—or perhaps a sense like a snake, for instance, which has a special receptor organ (a pit between the eyes) that can perceive nearby creatures by means of the heat they give off. Before visiting the Plutonians, therefore, we should have to devise instruments that would duplicate their electric-field sense and transduce the information thus gained into a modality perceptible to us.

What could such creatures as our hypothetical Plutonians look like? Since their primary sense is moderated by an electric-field organ that encircles their bodies, they are unlikely to wear clothing, and this means that their skin must be tough or resilient, either scaly, leathery, or shelled. A shelled exterior is very limiting, and on earth no highly intelligent forms have emerged from creatures encased in a chitinous exterior, so we must choose between a scaly or a leathery exterior to guide our imaginations.

On earth, scaly creatures have remained for the most part on the lower rungs of evolutionary development, so we are inclined to opt for the latter. As a matter of fact, among human beings a congenital skin condition, called ichthyosis, sometimes appears. It is a thickening of the skin that is reminiscent of our reptilian precursors, and since we know this condition in our own kind we recognize it as a potential physical character of intelligent beings.

Of course, an absence of clothing does not rule out body ornamentation and this is likely to be present, because indications of rank and role play a fundamental part in the emergence of intelligent life. Hierarchical organization of living groups is basic to evolutionary processes, and in higher forms there are usually perceptible symbols of rank. However, intelligence usually breeds individuality, and one of the ways for individuality to be asserted is by means of significant or symbolic personal ornament.

Now what kind of body decoration is likely to make sense to these heat- and electricity-oriented beings? Perhaps they have fashioned slivers of mildly radioactive rocks so as to create interference configuration: like those prismatic patterns that are set up in glass when it is put under pressure. Of course, we, as visitors, could not perceive those patterns with our kind of eyesight. All we would see would be the slivers of rock attached to different parts of their bodies, apparently by individual whim. But as soon as we referred to the instruments accompanying us to discover how these appear to the Plutonians, we should discover the jewel-like beauty of the rainbow hues intersecting in diverse geometric shapes. But the Plutonians would not be aware, as we should be, of the wonderful patterns of color and light revealed to us by this transduction. Their aesthetic pleasure would more likely be grounded in their perception of the forms of the radioactive emanations.

What would be the probable shape of their bodies? As we have said, there are certain prerequisites for the development of intelligent life of a high order, and these must limit our imaginations. First of all, because of the inherent chemical arrangement of replicating organic molecules, organisms tend to symmetry. Second, a highly evolved nervous system must have an organizing center, a brain, that must be protectively housed in that part of the body which receives sense impressions first. Third, there must be organs capable of grasping and manipulating. Fourth, there must be a means of locomotion.

Our Plutonians are unlikely to be flying creatures because of the nature of their atmosphere. If they are ground-living, they may propel themselves either by crawling like worms or snakes, or by raising themselves from the surface of the ground on limbs. Since the use of limbs facilitates a wider range of experience, an intelligent ground-living species is more likely to walk and run on limbs than to slither on a limbless body, especially as it is in any case going to require the modification of one or more of those limbs for grasping.

If limbs must be present, then how many shall we imagine? Nature is known to be parsimonious. Most of earth's higher creatures—if we exclude birds—are supported on four legs, two at each end. This arrangement, however, has not left them much leeway for manipulative skills, and most of them have been forced to use their mouths for grasping and carrying.

Some creatures, like the sea otter, have learned to float on their backs and use their forepaws as hands, and some others, like squirrels, kangaroos, and monkeys, may erect themselves on their haunches to free their upper limbs. The elephant, a very intelligent creature, has evolved an extension of its sensitive nose for manipulative purposes, and some monkeys, a specially developed tail. But we have to recognize in apes and, supremely, in man, the crucial importance of the hand in the development of intelligence.

Therefore our Plutonians will either propel themselves on four limbs and have one or two extra ones for manipulation, or they will have modified, as we have, at least two of their limbs, and perhaps some other bodily feature as well, for an equivalent of hands. Either result is possible.

If we are to choose, we are inclined toward the image of a

two-legged and two-handed being because that form ·is associated with intelligence on earth, but we have to keep it in
mind that a four-legged being is far more mobile, can cover
longer distances faster and thereby accumulate more experience of the external world than a two-legged one, and so a
four-legged creature with an elongated snout, tail, or, for that
matter, ears, remains a possibility. Which form they will have,
in that case, will depend upon whether speed and range are
useful in the environment in which the creature evolves.

Given the volcanic nature of the planet we have projected
as a home for beings informed by electric senses—and conse-
quently its necessarily rugged surface, pitted with craters,
scarred with chasms and cliffs, covered with boulders and
eroded mountains—we think that the speed of a four-legged
creature will not have been a factor in the adaptive survival of
the Plutonians, and so we shall endow them with only two.

Now let us consider their bodily shape. Do they have a
trunk with limbs at its four corners and a head at one end as our
higher ground-living animals do? Could the Plutonians be
spherical, a globe supported on mobile struts, with all their
organs on the surface or within the cavity of the sphere?

Could they, like octopuses, have a body with manipulative
and sensitive tentacles at one end, their mouths at the base of
their tentacles, and their eyes along with their electric organs
directly on the body surface? Our Plutonians would then have
to propel their headless bodies on their tentacles. But we feel
that the form of the octopus—by the way, also a very intelligent
creature and said to have an intelligence at least equal to if not
surpassing that of a dog—is supremely adapted to a waterborne
existence. Its body is supported by the water it lives in and so it
does not need a bony skeleton. On dry ground, in the long run, it
would need such a skeleton to anchor and support its anti-
gravity musculature. Although it remains a possible form, we
think it an unlikely one to have predominated in the rugged
terrain of our particular Pluto.

We might then ask ourselves what forms of land life other
than mammals have proved the most successful here on earth.
If we had visited our planet from space some millions of years

ago, we should have found it dominated by many species of reptiles, including the giant dinosaurs. Even among the dinosaurs there were some of moderate size and some that ran upright on two feet, but this seems not to have been enough.

The factor that was missing was the prolonged developmental stage in which immature young are nurtured and closely attended by care-giving adults, giving time and opportunity for brain function to develop exponentially. The dinosaurs, like contemporary reptiles, laid eggs, to which they probably gave minimal if any care. And in any case they coped with their environment by means of anatomical specializations of another sort: great size, thickened, armored skins, tusks, fangs, and claws. Even at the height of their dominance they never developed brainpower beyond its autonomic functions.

Much as we should like to imagine some wonderful form in every way different from our own for the intelligent denizens of other planets, then, biological considerations force us to conclude that while their size may differ from ours, in consonance with their planet's gravity, and details of their adaptive structure and sensory perceptions will surely be different from ours, yet overall they cannot be radically unlike us. The digestive functions demand an aperture for the intake of food (a mouth), a body with a cavity that can hold organs to break the food down, and an aperture through which the unusable excess can be eliminated.

There must also be a well-protected area to house the organizational center of the nervous system and the sense receptor organs, as well as information-transmitting mechanisms. On earth this area is invariably in the head. It would seem possible for these organs to be accommodated along with the digestive and circulatory systems in the trunk, even though on earth it has not happened that way. It could be that the capacities of a trunk are fully utilized in housing the digestive and circulatory systems while at the same time providing for the mobility of the organism. It looks as though we are forced to grant our Plutonians a head, a trunk, and at least four limbs, with only the possibility of additional limbs or specialized prehensile organs still left open.

As regards the face, the nature of the physiognomy would certainly be affected by the nature of the principal sense organ, as it is in all the higher living creatures we know of. Where hearing is vital, the ears show special adaptation—they swivel or are large; where a creature depends upon olfaction the nose protrudes or is projected forward in a snout.

Since the Plutonians are primarily served by electric sensory organs contained within channel-like depressions around their bodies, the features of their faces are likely to be unremarkable and small. The nose, sensitive to heat and only in a rudimentary way to odors, is more likely to be a depression above the mouth than a large feature, and the eyes will surely be small.

The head, on the other hand, is likely to be domed and large. To process electric information usefully, the brain must be at least as large as ours. High intelligence is a product of brain complexity, and there is not room in a very small brain for sufficient neuronal interconnections to produce our kind of intelligence.

On earth this begins to emerge when the brain approaches the size of those of the great apes, that is, about 600 cubic centimeters. In modern man, brain capacity ranges from approximately 1,200 to 1,800 cubic centimeters. If the brain of the electric Plutonians is to be housed in a head, as we decided it would probably have to be, then the head must accommodate a brain at least of this proportion in relation to their body size, should their bodies be bigger than ours; and they could hardly be much smaller than the smallest of our own species in order to support an intelligent-brain-carrying head. If they arose evolutionarily from a small species, natural selection would have favored its larger members, so that over time the size of the race would have increased.

The fact of intelligence again imposes certain limitations when we think about the probable life-style of our Plutonians. The development of intelligence, as we have noted, demands a long period of physical immaturity during which the young are closely protected. Sturdy homes capable of lasting many years are a necessity unless the climate is so benign that no elaborate housing is required—as, for instance, in the Pacific Islands, where a simple thatched shelter is sufficient, or in hot deserts where all that is needed is protection from the sun, which a tent provides.

We have decided that the planet of the Plutonians must have an unaccommodating, turbulent atmosphere in order for the electric senses to have taken precedence. Therefore it is unlikely that they will have built houses and towers like ours. We think it is more probable that they will have carved their cities into cliffs and mountains, at first utilizing natural caves and passageways within them, passing through village stages when their towns must have looked much like the cliff dwellings of the Pueblo Indians, and then, and over time, have elaborated these into great cities, securely protected from the dark, eddying atmosphere, in the hearts of their mountain ranges.

The darkness of these cities would be no problem for the Plutonians, since their electric-field sense would register the presence of any objects that impinged upon it, and the object would then become apparent to them as it would be to us by seeing it. However, the nature of their awareness is likely to make them far more conscious of each other than of nonliving material things.

They would immediately be aware of each other's emotions, since their feelings would inevitably affect the electric impulses they could transmit. Thus the Plutonians would respond like olfactory beings to involuntary messages of sickness or health, affection or dislike, pleasure or distress, and all their gradations, as well as to voluntarily adjusted electric emissions.

Lifeless material objects would merely impinge upon their receptor senses to make their presence apparent, but living beings would be known intimately to each other, and, given the care-giving propensities established in the Plutonians by the necessity of protecting their young, all their fellow creatures would be objects of concern and interest to each of them. Governed by an electric sense, warm, concerned interpersonal relationships are likely to be the guiding principle of their lives rather than material objects and acquisition.

In spite of all this there would have to be a basis for rank and social order. If private wealth is eliminated as an element in hierarchy formation, then the hierarchy has to be based on personal qualities. In an intelligent race living in such circumstances, we would think that especially intelligent individuals with gifts in electronic engineering would be esteemed. These individuals would plan mountain excavations for city sites, for sites for experimental and applied technology, for plant and animal genetics and husbandry, and so on. They would be the natural leaders who would direct the energies of their fellows in the service and maintenance of their communities.

We think they might have devised electric equivalents of the laser beam to aid in cutting out their city areas. Some among them may have discovered that strange property of electricity that we call light, and could well be devoting experimental caverns to determining the effect of this property on plant growth for food purposes. We could imagine that such lighted areas would be physically uncomfortable for the Plutonians and that work in them would be thought of as dangerous and undesirable, a little in the way that we ourselves think about working in mines. The production of synthetic materials for use for tools and utensils would also be a field for the exercise of intelligence and leadership.

Communication among such beings would probably not be verbal; more likely, it would be carried on my means of modulations of electric waves. Naturally the recording of their thoughts and communications would have to be in symbolic forms analogous to our writing, but its nature would necessarily be completely alien to us.

It might be that they have developed some modified radioactive materials on which their symbolic communications could be impressed in a kind of bas-relief that would emit differential radioactivity that they could "read." It is also possible that some of their most gifted intellectuals could have developed a way of scanning magnetic tapes directly, without necessity for transducing instruments. Whatever form their communication took, we would require not only a battery of transducing equipment, but also a great deal of inspired study before we could have any hope at all of gleaning their meaning.

On the other side, if we should ever have any contact with such beings, we should be totally incomprehensible to them. We should be apparent to them in the way that nonliving things are. Their electric field would register our presence and our appearance, yet to them we should be moving, apparently living objects that did not emit our own electric field. They would have to assume at first that we had turned off our emissions in order to hide from them, and so they would meet us with a good deal of suspicion.

Knowing this, we should have to carry equipment that would produce electric-field activity artificially for us. Being ignorant of how to use this with them communicatively, however, the best we could do would be to broadcast some sort of regular pulse in the hope that they would recognize this as an effort of communication. It would sound to them positively primitive, like children randomly banging on drums in contrast to the communication of music, and at first we should certainly seem to them to be barbaric and uneducated creatures, although, given time, we might well discover a method for mutual comprehension by electric means.

Even if we were able to do so, however, they would probably find our mission incredible. We imagine that their type of atmosphere would preclude the evolution of winged forms of life, and would also militate against their searching outward from their planet. Without birds and flying insects as models, their thoughts would hardly have turned toward space exploration. They are far more likely to have concentrated their ideas and their ingenuity toward burrowing inward on their planet than to exploring outward from it. Their deities would be gods of volcanos and not of the heavens.

Although their electric sensitivity might well have made them aware of the presence of celestial bodies, it seems unlikely to us that they would have given much effort to investigating their nature, and so astronomy, a science in which many of our own emerging civilizations achieved great heights, would not be among their accomplishments.

In the matter of transportation, the ruggedness of their planet's surface, in combination with its uninviting atmosphere, is likely to have discouraged any but the most elementary type. They might well possess nothing more elaborate then conveyor-belt walkways to speed their progress along their tunneled interior passages and perhaps nothing more than domesticated four-footed and fleeter animals than they for travel on the surface.

For the rest, as intelligent creatures, our Plutonians would be playful; they would certainly have devised some forms of entertainment for themselves, since it is not possible for a highly organized brain not to exercise its function at almost all times. Also, being a care-giving species, they would certainly have evolved some kinds of social and affectional bonds and the emotions that accompany these.

But their thought processes and therefore their personal needs in this respect would surely be so different from our own that it would be practically impossible for us to imagine what these bonds and feelings might be before meeting them—any more than we could have invented the responses that hold together a community of bees, and imagined ourselves into their bodies and feelings, had we no living examples before us for observation.

We could only prepare ourselves to look for some patterns of ties and relationships between individuals and of pleasureful entertainment, and hope that we could recognize them for what they were when we met them. There are so many almost incredible aspects of life here on earth—forms and functions that we could never have dreamed of had not life formulated them and presented them to us as models—so we can only remain open-minded about the possibilities of other planetary conditions in these respects.

Having said all this, we must now ask ourselves whether our planet of the Plutonians is the only possible ambience for intelligent beings that have evolved electric senses. We are obliged to admit that it is not. If we go right back to our original premise that the first essential for the emergence of such beings is that their environment is dark, we cannot entirely eliminate the possibility that they might exist at the bottom of deep oceans.

The World's Fair held in New York shortly before the Second World War familiarized the thousands of visitors who saw it with a most convincing model of a city on the ocean floor. The difficulty, as we see it, is how a very high type of intelligence could have emerged in that ambience. The Fair's model was projected as a place to which surface creatures could emigrate rather than as an extension of an evolutionary development of indigenous life. But let us nevertheless turn our imaginations toward this possibility.

When we first thought of the ocean depths as a possible environment for our electric-sensed beings, we were inclined to dismiss it because the necessities of adaptation to a watery world would include the streamlining of bodies for swimming, a factor which has modified the limbs of most of earth's mammals that have taken to the sea as a way of life. The sea otter is the only one we can think of that has retained hands, but it lives at or so near the surface that one could not imagine circumstances that would tend to the evolution of electric senses in such an animal.

Electric senses arise on earth only in very deep water, unless some special circumstance, like turbulence, makes other senses useless in shallower ones. Moreover, the electric emission and perception of the ocean depths is of a different kind from the sense we have been describing as a basis for life for our Plutonians. It does not exclude vision. On the contrary, it seems to have evolved as an aid to vision—that is, it makes visible those creatures that generate it in regions where light from the sun no longer penetrates.

William Beebe, the deep-sea explorer, reported in *Half Mile Down* that he first came upon animal illumination at a depth of 600 feet, and that at 2,000 feet there were never less than ten or more lights—pale yellow and pale bluish—in sight at any one time. Fifty feet below I saw another pyrotechnic network, this time, at a conservative estimate, covering an extent of two or three feet.... Another hundred feet and Mr. Barton [his companion] saw two lights blinking on and off, obviously under the control of the fish.... From 2,050 to 2,150 feet I saw relatively few illuminated organisms but later, at 2,200 feet, the lights were bewildering.

But if we are going to meet the requirements necessary for the emergence of *intelligent* beings, we need some ambience where deep water is not the only factor. It must be a region where limbs, or at least hands, are useful, and one that presents a challenge to manipulation. It must also offer opportunities for the prolonged protection of slow-developing young. As we are writing these specifications we cannot help thinking of the possibilities of the vast canyons that rift the midocean floor.

Suppose some other planet were to contain such features, deep canyons with cliff walls, caverns, and a variegated terrain, deep enough below the water for no surface light to reach it but not so deep that the water pressure might preclude the development of life forms large and complex enough to have evolved reasoning brains. Suppose we had started on this journey when we imagined such beings with electric senses: where would it have taken us?

The planet of the Hydronians would be covered with water. It might, here and there, show some islands where the tallest mountain peaks rose above the surface, but they would be small, few, and far between. Nevertheless, it might be just these islands that provide the clue (in the form of the debris of artifacts) to visiting space travelers that intelligent creatures existed there, and might encourage the search for them.

Our Hydronians might well have bodies like mermaids, with the trunk and lower part streamlined for swimming, but with a head, arms, and hands at the forepart. Such beings have been imagined by our mariners since ancient times, and perhaps they do indeed exist somewhere in the cosmos on some other planet. Our own planet has produced many species of marine mammalians—the whales and dolphins, the walruses and seals, sea lions and otters, manatees, dugongs—and it is certainly possible that similar creatures have evolved elsewhere.

We think we cannot do better in providing a setting for the world of the Hydronians than by quoting Beebe's description of his first impressions of life at depths of 2,200 feet. He wrote:

> Pteropods were close at hand and a host of unidentifiable organisms. I would focus on some creature, and just as its outlines began to be distinct to my retina, some brilliant, animated comet or constellation would rush across the small arc of my submarine heaven and every sense would be distracted, and my eyes would involuntarily shift to this new wonder.
>
> ... The jet blackness of the water was broken only by sparks and flashes and steadily glowing lamps of appreciable diameter, varied in color and of infinite variety as regards size and juxtaposition.... Now and then, when lights were thickest, and the watery space before me seemed teeming with life, my eyes peered into the distance beyond them, and I thought of the lightless creatures forever invisible to me, those with eyes which depended for guidance through life upon the glow from the lamps of other organisms, and, strangest of all, the inhabitants of the deeper part of the ocean, those blind from birth to death, whose sole assistance, to food, to mates, and from enemies, were cunning sense organs in the skin, or long, tendril-like rays of their fins.

What circumstance of marine environment is likely to have produced the Hydronians? We could imagine that a need to find shelter for the protection of their young might have driven their forerunners to seek it in the chasms of the submarine canyons. There, over time, the development of prehensile forelimbs might have been facilitated by a life-preserving need to scoop out hollows within those caverns and, eventually, to buttress and enclose them. Quite possibly, larger predatory creatures may have posed a threat to their existence and forced this behavior upon them.

As their forelimbs became essential to their survival, natural selection, and probably sexual selection also, would have tended to refine those features until they not only became indispensable but also, through increased use, came to modify their existence. At the same time, elaborations of bioluminescence patterns would certainly have occurred as an extremely adaptable and useful means of identification.

Before we begin to extrapolate from this, let us first take a look at those creatures on earth that have evolved light-producing organs and see how they have used them.

In the depths of the sea the outlines of the bodily shapes of luminous fishes may be made out at very close quarters, partly by the gleam of the fish's own light; but at a mere few yards' distance, only the "constellation" of their light organs can be seen. Depths-living creatures can distinguish predator from prey and fellow species members from alien breeds only by means of the patterns and colors of the lights shown by the natural "lamps" every luminous fish displays in a characteristic way.

As Beebe described them, one species (Melanostomiatids) has twenty pale-blue lights in a straight row along each side, looking like lighted portholes on an ocean liner. It also trails a 3-foot tentacle with two reddish lights at the end of it, the rear one tipped with blue lights. Its powerful jaws are illuminated from the inside and must be stunningly alarming to the unlucky creatures that approach the gleaming baits.

The deep-sea hatchetfish sails under less dramatic colors: its tooth-shaped phosphorescent stripes glow eerily on its sides like the jaws of a human skull. A species Beebe named the Five-lined Constellation Fish has five arcs of beautiful lights along each side of its body, one curving along its "equator" with two arching above and two below it. Every line consists of large, pale-yellow lights, each surrounded by a semicircle of very small, intensely purple photophores.

The number of "light uniforms" is great. In the lantern fish family alone, about 150 species may be distinguished by the number and arrangement of their luminous organs. Sometimes the differences are subtle. Two genera vary only by differences in the number of lights just in front of the tail fin, one carrying but two and the other from three to six. Some differences that separate species are even smaller, "but [according to Beebe] the fishes recognize them exactly."

Not only do species identify themselves by their light patterns, but within species males also often differ from females in the lights they display, much as male and female birds develop different plumage. The female of one lantern fish carries from three to five luminous plaques on the underside of the root of her tail, while the male has one to three light points on the top of it. In fact, the vision of deep-sea fishes, living in perpetual darkness, is finely attuned to "registering and interpreting abstract but precisely characteristic light patterns."

Many fishes have specialized beyond the use of their light equipment for identification alone. Some use special lamps attached to tentacles to attract prey; these remind us for all the world of baited fishing lines—which, in effect, they are. Moreover, their use is versatile. Some dangle lights that look like luminous marine worms just in front of their mouths. Some dangle their lights only until their pressure-sensitive lateral line organ informs them of an approaching creature, then they retract their "rod" and attack the intruder.

The barbel fish have "beards" rather than "rods," with lighted protuberances at the ends. In one species the fish measures a mere 11 inches, but the "beard" is over 3 feet long. Nerves in the "hair" of their "beards" are sensitive to the approach of prey. A species of the viperfish family (found at depths between 1,500 and 7,000 feet) illuminate the insides of their mouths with 350 dots of light, attracting small fishes and crustaceans to swim right into them.

In some other species glaring lights are used as a defense to dazzle enemies, as automobile headlights dazzle human beings, especially on dark, rainy nights. When Beebe put his luminous watch dial in front of a lantern fish, the creature expressed its alarm with a series of flashes. He recorded one experience in the following words:

> I watched one gorgeous light as big as a ten-cent piece coming steadily towards me, until, without the slightest warning, it seemed to explode, so that I jerked my head backwards away from the window. What had happened was that the organism had struck against the outer surface of the glass and was stimulated to a hundred brilliant points instead of one. Instead of all these vanishing as does correspondingly excited phosphorescence at the surface, every light persisted strongly, as the creature writhed and twisted to the left, still glowing, and vanished without my being able to tell even its phylum.

A deep-sea shrimp of the genus *Acanthephyra,* in moments of excitement, releases "a veritable rain of sparks from its arsenal," enveloping its enemy in a blinding and confusing storm of light.

H. W. Lissman, in 1959, proposed the theory that, for these fishes, flashing their lights served as a means of communication. He believed that besides the characteristic arrangements and colors of the light patterns, there were also equally characteristic signal codes, of the type we ourselves use in lighthouses. He thought the phenomenon represented a kind of Morse message by which the creatures in the depths could summon mates, warn rivals, and perhaps exchange information.

This idea found some confirmation when the Japanese biologist Terao found that the body of the luminous shrimp glows with 150 light dots that can be turned on or off almost instantaneously. Within one or two seconds green-and-yellow light patterns flicker from head to tail in quick succession.

It sometimes seems strange to us that so many of the devices mankind has constructed by the use of our brains and hands, and that we have believed to be monuments to our ingenuity, find their counterparts in nature, formed by biological means. But if we consider that the raw materials available for manipulation by us and by natural forces are the same, then perhaps we shouldn't be surprised that quite often our solutions are similar.

Consider, for example, the equipment of the luminous squid, a creature whose excellent intelligence is on a par with the octopus's, and which often hunts cooperatively in schools. Lorus Milne and Fritz Bolle have described their highly developed luminous organs, which have lenses, concave mirrors, diaphragms, and shutters that one is driven to compare with man-made searchlights. They are moved about with muscles so as to throw their light backward and downward, the direction of the squid's movement, so that they light its way, illuminate its prey, and dazzle its enemies.

The "searchlight" apparatus is obviously immensely useful to deepwater creatures. This is underlined by the fact that two types have evolved, open ones and closed. The open luminous organs mostly appear in shallower seas and do not make their own light. For this service, they depend upon luminous bacteria, which they collect from the sea water into small sacs under their skin or into clusters of tiny tubes. The microbes are attracted to the sacs and held in them by means of a nutrient culture medium produced by the squid's body processes and stored within them.

The closed luminous organs of the creatures of deeper waters, however, produce their own light by means of glands that secrete a luminous liquid. Almost incredibly, as Hans-Eckhard Gruner has reported, they can produce the most varied colors. He wrote: "By different devices like colored filters (skin pigments) and glittering mirrors, all sorts of color shades may appear in one particular creature. It may produce red, blue, green or white light. Its luminous organs, when active, will then sparkle in these colors like precious stones." One species, the Infernal Vampire Squid, which has its tentacles joined by a membrane like a bat's wing or a duck's feet, carries reflectors on stalks at the end of its body. If it wants to "turn off its lights," it retracts these stalks into pockets, which it can close with the equivalent of a lid. Some squids that cannot switch off their lamps have various kinds of shutters that they can introduce between the source of light and the lens to effect a sort of natural blind, or blackout curtain.

The more we discover, the more we are amazed at the seemingly boundless variety of organs and responses possible to living matter, any one of which conceivably could give rise to an intelligence if it met the necessary circumstances and were placed in the necessary environment. With this information about some of the earth's luminescent species, we may now turn our thoughts back to the Hydronians.

The first thing that strikes us about the perceptions of the deep-ocean creatures is that they recognize each other by abstract signs—we recall that they recognize each other only by the patterns of light that their bodies display and that are characteristic of certain groups—and that this form of information transference lends itself extremely well to sophisticated development and use. Moreover, the ability to turn on and off such light displays, or parts of them, in different time sequences, may form a basis for signal codes quite as useful as the letters of any written language. We can also imagine insignia, uniforms, or other kinds of body ornamentation provided by little clusters of luminescent bacteria tucked into the skin sacs like the squid's.

For other examples of this type of communication on earth we do not have to go to the ocean depths. When the courting black firefly of the southern United States seeks a mate, the male emits a flash of light lasting precisely 0.06 seconds every 5.7 seconds. When a flashing male approaches within about a dozen feet of a female, until then imperceptible in the grass, she responds exactly 2.1 seconds after each flash by blinking back. The suitor then approaches her; not more than approximately half a dozen of these exchanges are necessary for their purpose to be consummated. A researcher, testing these responses artificially with a lamp, discovered that he had to time his flashes with extreme accuracy for the invitation to be understood. If he blinked it by only a fraction of a second off time, the firefly ignored him. This is an impressive illustration of the precision of this type of communication, and in our world many kinds of insects as well as of fish use it.

Among them are the Central American beetles we mentioned in passing earlier. Their darkness is the inky black night of a tropical rain forest. From his observation station at Rincón on the Golfo Dulce in the rain forest of Costa Rica, a naturalist named Charles Hogue reported,

> There's a beetle here of the genus *Pyrophorus* which has two headlights on the front of the thorax and another light organ on the underside of the abdomen. . . . They're not like fireflies that flash on and off. They glow in a steady stream of light which flows through the forest. . . . The light is so bright that it actually lights up the leaves beside the beetle as it flies . . . [with] a greenish fluorescent light. . . . There's one that was bright, bright orange like . . . a burning match or a sparkler thrown through the air. One could easily use these beetles to live by, to read by. They're like meteorites falling through the forest.

He also wrote of other forms of luminescent life:

> There are bacteria and fungi which are responsible in part for the decay process of dead wood, and these produce luminescence. . . . You can observe this soft, greenish, cool, ghostlike light emanating from logs and branches in the forest, and in the pitch of night it's almost like a soft, ghostly picture painted in the darkness.

Clearly, the floor of a rain forest is also an ambience that must be considered as a possible world engendering electric senses.

We were inclined at first to think that bioluminescence of itself could not have been a sole channel of information for a species of high intelligence, but we now discover that it might well be a contributory one. Other senses would have to add their quota and support the light organs, much as speech, hearing, smell, and touch support our own primarily visual intelligence. Sight and hearing would surely be among the faculties of the Hydronians, too, and in those of them that are marine creatures, probably also the lateral line sense. We would also imagine that some vocal resources, perhaps of the types used by our dolphins, would have developed by convergent evolutionary processes in the Hydronians.

The remarkable thing about the large cetacean brain is that its two hemispheres often work separately and, what is more, rest separately. When a dolphin sleeps, one of its eyes is always open and alert, as is the brain hemisphere to which it reports. In addition, the dolphins have two distinct means of vocal communication. The first is a vocabulary of whistle-like sounds, which they produce all the time when they are in each other's company; they also exchange trains of clicking sounds.

Dr. John C. Lilly, who has conducted extensive research about the dolphin mind, showed that the two kinds of sonic exchanges do not correspond in time, that is,

> they can be talking with whistles and talking with click trains, the whistles and clicks completely out of phase with each other. . . . These observations led to further studies in which we demonstrated unequivocally that each dolphin has at least two communication emitters, both in the nose, i.e. below the blowhole, one on each side. . . . Thus a given dolphin can carry on a whistle conversation with his right side and a clicking conversation with his left side and do the two quite independently with the two halves of his brain.

A double communications ability of this sort, if combined with manual dexterity in an intelligent species, leaves one's mind agog with thoughts of its possibilities.

The technology developed by the Hydronians may have been limited to what is necessary for food and shelter, but, given the precision of their sense modalities, they may well have explored and turned to advantage many marine sciences. They may have learned to use the diverse potentials of marine bacteria, including cold light, or have discovered submarine oil deposits and improved the quality of their lives with oil's chemical by-products. Surely they would have developed fully all the potentialities of electricity, perhaps using it to separate water into its component hydrogen and oxygen, which they could then use for cutting and boring and for melting the various minerals they might mine.

In filling out our ideas about the Hydronians, a problem that presents itself is the matter of records. For elaborate technologies or, indeed, any high culture to arise, there must be a means of recording and passing on information. In a deep-sea habitat, writings on paper or parchment are unlikely. They would be technically possible if an air-breathing species had migrated to an ocean floor and founded cities by blocking off the surrounding waters and producing an artificially dry atmosphere, but a species that had evolved in the depths would be unlikely to have thought of devising dry areas for record storage.

We believe that if, anywhere in the cosmos, an advanced species with an intelligence on a par with ours has evolved in a watery environment, their stores of knowledge must be preserved in the phenomenal memories of a special caste and conveyed from generation to generation by an oral tradition. We would have two paradigms for this assumption. The first is the amazingly complex brain of the cetaceans, and the second is the simple fact that until comparatively recent times—that is, until something like five thousand years ago—the memory of our own old people was mankind's chief if not sole repository of knowledge. The memory of our elders remained a factor in the evolution and preservation of culture until our own time, when universal education has brought the ability to use the written record within the scope of populations at large.

As regards the cetacean brain, Dr. Lilly has written that

> the sperm whale's brain is so large that he only needs a small fraction of it for use in computations for his survival. . . . To think the way we do he would need to use about one-sixth of his total brain. To him, our best thinking may appear to be reflexes, automatic and primitive. The rest of this huge computer is computing continuous inner experience beyond our present understanding. If a sperm whale wants to see-hear-feel any

past experience, his huge computer can reprogram it and run it off again . . . [it] gives him a reliving, as if with a three-dimensional sound-color-taste-emotion-re-experiencing motion picture. He can thus review the experience as it originally happened. He can imagine changing it to do a better job next time he encounters such an experience. He can set up a model of the way he would like to run it the next time, reprogram his computer, run it off, and see how well it works.

Of course, we can make similar calculations by means of our artificial computers, but to be able to do it all in our heads would enormously simplify the process and present vast possibilities. On this basis, then, we see now that we need have had no hesitation in imagining a brain evolved in an ocean environment of some other world that would have a potential at least equal to ours.

The idea of the knowledge of a species being preserved and passed on by a specially evolved caste has as a model mankind's own evolutionary history.

We have referred to the fact that a prolongation of infancy and childhood is an important factor in the development of an intelligent species. This prolongation is biologically achieved by a tendency in that species toward a slowing down of all its developmental phases; that is, there is natural selection for rate genes (those genes that determine the rate of growth and development) that effect a *slower* rate of development in their case. Of course, when this occurs, not only childhood but also all the other phases of life are prolonged and, as a result, intelligent species are inevitably longer-lived than their less intelligent evolutionary progenitors.

This process creates a caste of postmature, elderly individuals for which a natural breeding group would previously have had no function. The function of immature creatures is to develop and learn. The function of the mature is to breed and to sustain life. It is a cliché, but nonetheless true, that nature does not tolerate a vacuum: where a feature arises, use is made of it; where a function becomes necessary, or useful, an organ is adapted to perform it. Thus, as a postmature class evolved in mankind, a function was waiting for it to fulfill. Mankind's elders aided their species' survival by carrying knowledge of our lore and skills in their heads and thus became focal points, aiding the cohesiveness and survival of their societies.

Today, we still carry biological signposts to this path of our progress in the structure of our brains. The human brain has different centers for long-term and short-term memory. The interesting fact is that as an older person's physical prowess begins to wane, the short-term memory wanes with it. But the long-term memory, the faculty that made the elder a vital factor in the development of the human group, still persists in the old person to the very end of life unless the brain sustains damage. For this reason, our old people's memories are usually remarkably clear about the events of their early lives, but sometimes dim about recent and current ones.

On this basis we would postulate that our Hydronians live to very old age, that they receive honor and respect from their juniors, and that they perform the specialized function of re-membering the body of tradition and skill acquired by their fellows, adding the store of each new generation, and passing the whole on to the rising generation orally. As the fund of knowledge increased, there may have developed a degree of specialization among the elders, each committing to memory the knowledge of a certain field—a specialization that might have become the tradition of certain families, or lines.

How would this oral tradition be passed on?

If we again turn to mankind's experience for inspiration, we are impressed with the fact that wherever tribal history is passed on by word of mouth, this is usually done in the form of epic sagas chanted on ritual occasions. It seems that the rhythms of poetry and music make it easier for the teller to remember and the listener to learn the tale. Not only in tribal songfests but also in our own practices we find elements of this in the chants of the medieval Christian church and in the chanted liturgies of many other religions; in the fact that epic poetry is the earliest form of the literature of modern lan-guages; and in the rhymes and nursery songs of childhood.

Moreover, chanting or singing would tie in well with what

we know of the communication of whales, and tone quality is also a factor in much human vocal communication. In the Chinese language, for instance, the same syllables expressed in different tones or accented differently sometimes have different meanings. Melodies are capable of almost infinite modulation, and so we would think that a race dependent on memory storage and the dissemination of knowledge by sound would probably "sing" their vocal communications rather than "speak" them. Also, where communication of this type is used, we may imagine a highly ceremonial kind of life for our Hydronians.

On the other hand, it is possible that a deep-sea intelligence aided by bioluminescence may have taken a different turn altogether. Light may be modulated in several ways. Among them are intensity, color, the length of time separating flashes, and the position in which lights are arranged. Therefore it is quite within the bounds of possibility that bioluminescence itself might be tied to a system of communication so sophisticated that oral and aural means remained rudimentary. And this would promote a completely different way of understanding and thinking and a totally other set of life patterns.

We opened this chapter with an invitation to our reader to speculate along with us. Having hypothesized our Plutonians and our Hydronians, as you see, the subject is still wide open.

SEVEN

Other
Ways of Knowing

WHEN IT COMES TO serious speculation about whether life has arisen in other worlds, on the whole, the scientist concerned with astronomy or physics has been our authority. Using whatever data he has been able to provide, the science-minded lay person has taken it from there and allowed imagination to play with the few, albeit increasingly numerous, hard facts available.

It would seem that physics and the mechanical sciences have been more attractive than the biological sciences to the imagination of the science-minded lay writer, because many of the speculative descriptions of science fiction display their authors' wide knowledge of physics and mechanics, but usually have very little basis in anything that we know of the processes of living organisms.

Scientists working in the biological disciplines are not much given to speculation. Life is varied and at times incredible enough, and those who seek to understand its ways are concerned with well-documented facts about it. Among physicists, on the other hand, informed but imaginative speculation is at a premium. Over and over again, a concept arrived at purely theoretically through leaps of creative mathematical imagination has later turned out to be fact, confirmed by empirical evidence.

In this chapter we are going to offer some biological "raw material" for imagination by describing some other kinds of intelligences that are hard facts—they exist. May we suggest to our readers that they use these facts as starting points for imaginative explorations, for the discovery of other worlds. It might well turn out that ideas formulated in this way could, by chance, pave the way to knowledge of extraterrestrial life as, somewhere, it really exists.

As we collect raw material for our mental construction of other kinds of being, we must separate it into two categories. The first comprises the possibilities of the senses that, via their messages to the brain, inform a being of the nature of the world around it. This category includes what happens to the message within the brain that has to process the information it receives and convert it into symbols that can be used by the being that receives them.

But the information brought to the brain by the senses and the brain's elaboration of those messages is only half the matter. An ability to communicate our needs and to pass on to others what we know is the other half. Intelligence, to us, means the function of the human brain. But there are many other kinds of brain all around us, collecting, relaying, and using information all the time.

These systems, too, should be recognized as alternative intelligences—other ways of knowing. The method by which information is received, the organs that assimilate it, and those that execute responses to it, must all affect the way a creature perceives and responds to the world about it, confining its understanding within the limits of its own perceptions and its behavior in accordance with its understanding.

Here, now, are descriptions of some aspects of some of the very many senses with which we have not dealt in any detail so far.

ECHOLOCATION

Echolocation is another of those phenomena that mankind has discovered so recently and acclaimed as a product of scientific perspicacity, only to find that nature had been using the properties of sound waves for information and communication ages before our species emerged from its simian origins.

Essentially, hearing is derived from the perception of vibrations, and different species perceive vibrations at different ranges. Physically, vibrations are rhythmic compressions of air that may be perceived as they reverberate against a drumlike arrangement of tissue—in man, of course, the eardrum.

Sound is generated only by vibrating bodies, and the wave motions they set up must be transmitted through a medium. For us air is usually the medium, but any other gases, liquids, or solids may conduct sound. Sound will not travel in a vacuum, and this is a factor to have in mind when one thinks in terms of space travel. If we place a ringing bell into a bell jar and gradually remove the air, we find the sound gradually diminishing until we cease to hear it. If we replace the air, the sound returns.

The nature of the medium through which sound travels affects the speed of its movement from transmitter to receiver: the denser the medium, the faster the transmission. In air at a temperature of 15°C, sound travels at 1,100 feet per second, but in water its velocity is four times as great, and this speed is quadrupled again when the medium is a solid of such kinds as steel, glass, rubber, or the hardest wood. (Here is a modulation that could be significant for conveying information to an acoustically oriented intelligent being.)

In point of fact, sound, being a vibration, can be *felt* by the skin, bones, or any other part of the body, but only when a receptor organ such as an ear is present can it be *heard*. The capacity to sense vibrations is one of the functions of the peripheral nervous system, and in many animals is an important one, although it is not normally brought into use in man. However, if a tuning fork is placed against, say, the shin bone, vibrations can be sensed. This affords a useful medical tool in certain diseases of the peripheral nerves, such as those that arise in the wake of pernicious anemia, because sensitivity to vibration is the first function to go. This makes an early diagnosis possible. Conduction of sound waves through the bones can also serve as a diagnostic tool in determining whether deafness is due to a disease of the middle ear or of the auditory nerve. A vibrating tuning fork placed behind the ear will be heard so long as the auditory nerve is intact, irrespective of the health or disease of the middle ear.

Echolocation is only one of the many ways in which organisms use acoustic properties both for information and for communication. Creatures as vastly different as dolphins, bats, and night moths share this sense. In man, too, rudimentary elements of it exist. Some blind people tapping objects around them with a stick are able to gauge by the echoed sound of the taps their distance from a wall, a curbstone, or an obstacle in their pathway. Some show superior sensitivity to this kind of information. It is said that they have a gift for it. Unfortunately, most human beings, including the blind, are relatively impervious to acoustical information of this type and command only the coarsest perception of it.

We have already mentioned something about the dolphins' (and possibly other marine mammals') supersonic tonal communication and reception, but there are further aspects we haven't yet touched on. For example, when W. N. Kellogg described their "sonar" sense in *Porpoises and Sonar,* he said that the proportions and locations of fixed and moving things— rocks, great plants, fish, fellow dolphins—seemed to be perceptible to them by a form of hearing that we do not possess.

In discussing the detector function of the "sonar" and its capacity to distinguish between different substances, he suggested that the way such distinction is accomplished could be understood better by comparing it to vision or to optics. Daylight or white light, he pointed out, contains all the wavelengths of the, to us, visible spectrum. Yet when white light is used to illuminate a red surface, red light is all that is reflected back. The coefficients of "reflection" of the various surfaces are not the same.

Kellogg further suggested that a series of porpoise clicks is like "white noise." He thought that some of the original frequencies transmitted by the animal are absorbed and some reflected, so that the echoes from different materials would differ in composition or in quality, that is, that they would vary in the pattern of frequencies they contain.

"Wood, in other words, simply 'sounds different' from metal to a porpoise," he wrote, "in the same way that it looks different to the human eye. It is the sound spectrum of the returning vibrations which gives the clue to the nature of the reflecting surface."

Lilly, too, has given yet another slant on the dolphins' aural perception. He asked the question: "If we were placed underwater and looked at one another by means of sonar, what would each of us look like to the other?" Since sound waves in water penetrate a body without much external reflection or absorption, Lilly pointed out, skin, muscle, and fat are essentially transparent to any sound waves coming through that medium. The internal reflections could come only from air-containing cavities and from bones. And so we should see a fuzzy outline of the whole body, but the bones and teeth within it would be fairly clearly delineated. The most sharply delineated objects are any gas-containing cavities. We would have a good view of portions of the gut tract, the air sinuses in the head, the mouth cavity, the larynx, the trachea, the bronchi, the bronchioles, the lungs, and any air trapped in or around the body and the clothing.

Living dolphinwise, we would have little need for external facial expressions. The truth of our stomachs would be immediately available to everyone else. In other words, anyone could tell whether we were sick or angry by the bubbles of air moving in our stomachs. The true state of our emotions could be read with ease.

Lilly added a caution:

Please notice [he wrote] that in the above descriptions of the sonic acoustic underwater world I use primarily visual language "to see by means of sound." Since we are more visual than we are acoustic this is necessary, using our current language. This language requirement is reflected in the construction of our nervous system. We have ten times their [the dolphins'] speed, their storage capacity, and their computation ability in the visual sphere; the dolphins have something of the same order of speed, storage capacity, and computation ability in the acoustic sphere. This, then, is one of the major differences between the minds of men and the minds of dolphins.

Dolphins, of course, are not the only creatures that use this type of communication. Bats, too, familiarize themselves with their surroundings primarily by sonic reverberations, but few of us realize the extraordinary refinements and precision of this type of perception. Donald R. Griffin found that the auditory center of the bat's midbrain can discriminate between sonic pulses separated by as little as one-thousandth of a second. Moreover, the higher the vibration rate of the sound, the more amenable it is to being beamed like a searchlight. The brown bat's acoustic emissions rise from 50,000 to 100,000 per second, which corresponds to a wavelength of from 3 to 6 millimeters. This enables it to perceive and recognize even tiny flies and mosquitoes in flight.

It is a still unsolved puzzle as to how bats, sometimes massed together in their millions and often in thousands, and immersed in a sea of sound created by their own kind, are able to pick out the returning echoes of their own sounds, even though it is true that each individual has its own distinctive rate of vibration. To put this into terms of human experience, it is as if literally many millions of finely differentiated shades and degrees of intensity of color were beamed at the eye both simultaneously and consecutively, while a single observer were able to pick out from them a particular shade of, say, yellow that is barely distinguishable from a multitude of other shades of yellow but that happens to be that observer's own personal signal.

In a somewhat similar though far less refined fashion, human beings do possess something slightly like the bat's ability to filter out significant sounds from a welter of noise. In a crowded room, for example, we can hear what the person we are speaking to is saying even though we are surrounded on all sides with people talking—the so-called cocktail-party effect. A mother may sleep soundly through all kinds of noise—traffic, or sirens, or thunderstorms—but awaken at the slightest whimper of her baby. However, neither of these examples displays the degree of refinement of the bat's discriminatory sonic abilities.

Again and again we are moved to wonder at the way in which nature's devices, honed by the selection of evolution, cover all kinds of contingencies. For example, it might be thought that when a bat is eating a mosquito or a fly, it would be temporarily unable to send out its vocal emission and would thus become liable to collisions. But, as Professor Anton Kolb discovered, the fact is that the bat has alternative methods for orienting itself at those times: it can either whistle through its teeth or, if the prey is too fat, it can emit signaling sounds through its nose.

Even at this degree of refinement, many variations are possible and do occur. The horseshoe bat, for instance, does not send short volleys of ultrasonic bursts, but rather series of longer-lasting, pure and undistorted ultrasonic beams; again, each bat sends at its own particular rate of vibration. The remarkable thing about this system is that the returning sound would be drowned by the continuing emission were it not for the bat's ears, which, as F. P. Moehres discovered, move up to sixty times a second, constantly scanning the environment point by point, apparently to distinguish the angle of the echo.

(A sense of this type, so extremely malleable both for perception and for communication, would provide an intriguing challenge to the reader embarking on a speculative journey to another world. As a prime mover in a creature's existence, modifying its anatomy, indicating the way of life that formed it, and as a basis for an advanced intelligence, it could form the keystone of a very elaborate imaginative edifice.)

We can hardly leave the echolocation sense without mentioning the night moth, for which information by echolocation is so fundamental a part of its world that it has evolved physical defenses against being perceived by other creatures like bats, which have a sense similar to its own. The night moth has evolved a wing structure that eliminates air turbulence—and, therefore, sound—around its edges as it flies. The biophysicist Heinrich Hertel discovered that this is achieved by fine fringes along the turbulent flow area of the wing. Of course, this does not entirely eliminate the moth's vulnerability to the bat, since the bat's echolocation apparatus can still detect its presence, but it does eliminate at least one hazard—the sound of its flight.

Night moths also possess ears sensitive to ultrasonic sounds. These ears, located on both sides of the thorax near the waist, enable the moth to sense the presence of a bat within a flying distance of a hundred feet—a considerable distance, given the size of the moth, and one that gives it time to drop, swerve, and maneuver to avoid being caught.

As we have indicated, a sense capable of such a high degree of precision as echolocation could well be an instrument of an evolving intelligence in some other corner of the universe. With the evolution of an intelligent being, a sonic sense of this kind could lend itself to exquisitely refined use.

In projecting what kind of being, environment, and culture might arise in these circumstances, many possibilities immediately present themselves. As in the dolphin, the echolocation capacity might be just one of several highly developed senses, in which case the intelligence of the being could conceivably be extraordinary. Or it might be an intelligent creature that flies and because of its fast motion needs the precision of such a sense.

Alternatively, or in addition, it might be a night-living being that sleeps in the heat and brightness of the day and emerges to constructive energy in the nighttime. Imagine a social setting in which a number of such creatures live and interact with each other. Even more challenging is an endeavor to create a scenario in which we come into contact with such creatures as these. What difficulties would arise and what hurdles to communication would have to be overcome?

We leave these questions to the reader to add to his own and form a picture of another kind of intelligence—and world—that it might reasonably be possible to expect to encounter elsewhere in the universe.

TACTILE SENSES

Information brought to a being's brain via a tactile sense is sometimes of a different nature from that brought by most other senses, in that the sense of touch may be active or passive. When something touches us—whether a leafy branch, another creature, a gust of wind, or the warm radiation of a fire—we feel it passively, or involuntarily, and our skin informs us of its presence and nature. But if we wish to find out about the shape or texture of a nearby object, we may also reach out actively and deliberately investigate its nature by touch.

Of course, an animal may smell something passively or deliberately seek out another by sniffing for it, but with the tactile senses this is more pronounced. Many tactile organs express in their form this searching quality of the sense; not only fingertips, but also whiskers, feelers, tentacles, and similar structures that "seek out" rather than "wait to receive" information monitored by this sense, project from the body, sometimes quite distantly.

The tactile sense is probably the most primitive of all the senses in that there is no living matter, however simple, that does not respond in some way to being touched. In this respect, tactile sensitivity is an adjunct to any other sense that may evolve. No complex creature, and certainly no intelligent one, can possibly evolve with one sense alone. To this extent a sense of touch surely accompanies and supports any and every other modality. Nevertheless, we see on earth some very remarkable refinements of tactile senses, and we can imagine circumstances in which some elaboration of these may emerge as the primary sense of an advanced intelligence.

In contrast to most tactile organs, which are primarily

proximity receptors, there is one that is a distance receptor: the lateral line organ of fishes. We have already described how bundles of nerves in these grooves register the differential pressures of the surrounding water and thus enable the fish to perceive objects within its range. We could imagine that such an organ might also serve an air-living creature by registering differential air pressures. Were such a creature intelligent, of course, it would have to have a way of transmitting as well as of receiving information. While in most species the modalities of emission and reception of information are not the same, one could imagine the possibility of highly sophisticated communication by means of the soundless emission of graduated air pressure.

By a mechanism along the lines of a whale's blowhole, perhaps, it might be possible to emit air by a very finely adjusted bellows action of lungs in a way that could convey meaning directly to an air-pressure-sensitive being. Our own awareness of differences in air pressure is crude. We feel the difference between a breath of air blown in our direction by a person's exhalation and a slight draft from an open door, between a gentle puff of wind or a light breeze and the sharp slap of a gale.

Were we to possess an organ capable of discriminating precisely between all the degrees of strength from a breath to a puff to a blast, we should have the elements of useful communication in this air-pressure factor. Lacking such a natural organ, we could to some extent duplicate it by using an exponentially refined anemometer—the instrument our meteorologists use to measure wind force.

The idea of a language borne on airflow is attractive. One can imagine nuances equal to or perhaps surpassing those of smell and surely exceeding those of the discrete sounds of speech. It could be a highly poetic or musical language, and provide a wonderful medium for the expression of emotions.

By contrast, direct touch onto the skin surface, another facet of the tactile sense, could provide a very precise and also versatile means of communication. A supreme example of the possibilities of this form of communication is available to us in the life of Helen Keller, who learned to interpret all human messages normally conveyed through sight, sound, and speech, through a system of direct touch. By this means, a system of symbols as precise as letters of the alphabet could be built up, using not only differential pressure but also the part of the body touched.

Our alphabet of twenty-six letters can easily be substituted by twenty-six touch points to gain an exact equivalent of our own language, and the number of those touch points could be increased as easily as we extend our alphabet by such symbols as plus and minus signs, punctuation marks, and so on. Whereas tactile communication by differential air pressure seems to be supremely suited to poetic expression, direct tactile communication of this sort would lend itself exactly as effectively as spoken words to the purposes of science, technology, and the precise arts.

In suggesting these possibilities for a cultural evolution based on tactile senses, it is useful to remember that wherever mankind has invented equipment, like radar or sonar, for the extension of our senses or to substitute for senses we do not possess, these tools are crude instruments when compared with their equivalents naturally evolved in living organisms. No matter how incredible an idea or an invention may seem, we can be sure that if in fact it has occurred somewhere in the universe by natural means, it will surely be more versatile, more refined, and more incredible than anything we can dream up based on our own experience.

The rich potential of an actual tactile sense as seen in nature is demonstrated by one of the many abilities of the bees. They have an absolute architectural sense for building hexagonal honeycomb cells, which they construct with undeviating accuracy within one-tenth of a millimeter. This ability is based upon the tactile messages of some of their numerous sensory bristles.

Other tactile organs found on earth that are of interest for a quite different reason are the pigmented spots of elevated tissue around the mouths and lower jaws of alligators. With these the alligators sense the presence of edible fish in the mud at the bottom of their rivers. It has been suggested that the hairy moles that not infrequently develop around or near the human mouth have their origin in this feature. That a hair, or several hairs—vestiges, perhaps, of sensory whiskers—often grow from these moles would tend to support the speculation.

Probably the highest intelligence that has evolved on earth primarily on the basis of a highly developed tactile sense (in this case in conjunction with excellent eyesight) is that of the octopus. Its entire body surface is extremely sensitive to touch, but its arms, with the rows of suckers along them, are its supreme tactile organs.

An octopus's arm is both prehensile and exploratory, and although it does not have fingers and an opposable thumb, it is a tool that is at least as useful and developed as a hand. It can both envelop or grasp an object, and also adhere to one with what has been called a "terrible efficiency." The tip of each arm is as delicately sensitive as an antenna; with it the octopus can handle the smallest morsels; it can excavate the last fraction of meat from the shell of a lobster through a single incision. It is also exceedingly strong; an octopus can drag up to twenty times its own weight, for instance, whereas a man can pull only double his.

All the octopus's arms are not the same. It uses the two arms in the axis of the eyes, called the dorsal arms, for feeling and grasping, and the next two also for working—seizing crabs, picking up objects, building its "home," and so on. An octopus, moreover, is able to coordinate different activities with different arms simultaneously—it can, for example, pick up a crab with one arm, at the same time warding off a dangerous object with another. (One has to recognize the complexity of a nervous system and brain that permits an animal to do this. Many human beings have difficulty in carrying out different actions at the same time with each of their two hands.)

The third arm, in the male octopus, is different anatomically from the others, since it is modified as a copulatory organ.

With it he first caresses (touches) the body of the female, and if she accepts his approach, he then penetrates her body cavity with it. The male also has two extra-large specialized sucker disks.

Observers have noted that the octopus's sense of touch is more highly developed than mankind's, and that it is associated with a chemical-gustatory sense that is hard for us to conceptualize—perhaps something like being aware of the smell, taste, and feel of an object through a single sense modality.

Besides this, the octopus has excellent eyesight. Its eyes are almost human—extraordinarily highly developed and quite exceptional for an invertebrate, with eyelids, irises, crystalline lenses, and retinas. They are mobile, can rise up like periscopes, and can be turned to look in different directions. Divers who have observed octopuses in their natural habitat have all been especially impressed with the lucidity and expressiveness of their eyes, which reveal the animal as alert and curious.

Life without a skeletal frame has some advantages that we could envy. Octopuses can stretch themselves thin, like rubber, and turn their eyes obliquely, so that even the largest of them can slide into barely visible crevices and openings. They can elongate their arms by stretching them out; even the head changes shape, when necessary. With this ability, and its strength, the octopus has been called the supreme escape artist.

Jacques Yves Cousteau has described how he placed an octopus in a tank on the deck of his research vessel, the *Calypso,* where it succeeded in raising and removing the cover he had weighted with several 20-pound weights to prevent its escape. "We saw its arms emerge one by one, then its eyes, then its body slide down onto the deck," he writes. "We did not try to stop it. The sea was only a few feet away, and we were already feeling guilty about having deprived an animal of its liberty—a splendid animal, made for freedom."

Octopuses live in "houses." They may use ready-made shelters like crevices in rocks, burrow under large stones, or live in human artifacts from sunken vessels or discarded into the water from the surface. They have been found domestically installed in amphoras from ancient wrecks and in discarded pots or tin cans. Alternatively, they may build their own houses, using rocks, pebbles, rusted cans, bottles, sandals, even old tires; anything they find on the sea floor is put to use. What is more, they are fastidious housekeepers. They keep their homes clean, using their arms to remove larger objects, and their funnel, like a hose, to clear out sand or mud with jets of water.

As might be expected of a creature with such a complex nervous system, the octopus is very emotional. Its emotions cannot be hidden, since they are expressed by color changes. From its "normal" light-brownish color, an octopus may blanch to white or turn through stages of pink to a deep, passionate red as it expresses fear, irritation, anger, excitement, or any other feeling.

With their fragile and sensitive bodies, they like to be petted. The divers and researchers who work with them say that they seem to "like" some people and to "dislike" others, and that they appear to enjoy cooperating with people if they are played with, petted, and stroked, and, above all, handled gently.

With all this, however, the chief impression made by the octopuses on those who have dealt with them is one of lively curiosity and intelligence. The Cousteau team conducted many experiments to test their ingenuity, and they passed them all with flying colors. Among the tests was one in which a live lobster was put into a large, clear, glass-stoppered jar filled with water. The jar was taken down to the sea floor by a diver and left in front of an octopus's house.

The diver reported that the octopus opened the discarded suntan bottle door of its abode and eyed the lobster for a moment. Then, suddenly, an arm shot out like a whip and grasped the jar—whereupon surprise paralyzed it for an instant. It seemed clear that the invisible wall between itself and its prey was something new in its experience. It changed color, turned red, as puzzlement, surprise, and anger affected its pigmentation. The octopus seemed to the diver to be thinking, looking—then deciding first to try the method that had previously always worked for it with crustaceans of the lobster's size.

Cousteau said that his diver reported that the octopus wrapped its arms around the jar, climbed on it, and covered it with its mantle. Under normal circumstances, it would now have been possible to paralyze its prey with poison from its salivary glands. Inside the jar the lobster began moving around, but this only increased the octopus's impatience. It then tried to take the jar into its house.

Failing in this, it then explored the jar slowly with one arm after another, feeling it from top to bottom. When the arm reached the cork stopper, it stopped and proceeded even more cautiously—until it found the small hole drilled through it. The tip of an arm was inserted through the hole and touched the lobster. The contact seemed to electrify both predator and prey, and the lobster gave a violent jerk with its tail. The octopus was reassured that the lobster was alive. Its respiration accelerated; it changed color constantly. Suddenly the cork was out. The octopus arm remained glued to it for an instant. Two other arms were already inside the jar, removing the lobster. The octopus raised the lobster to its mouth, immediately immobilized it, and then carried it to its home.

The observer Frédéric Dumas, noted,

> It was on the third try that the octopus whipped out the stopper as though it were something it had been doing all its life. When one thinks of how long it takes to teach a dog something as simple as sitting up or shaking hands, one must admit that an octopus learns very easily; and that, above all, it teaches itself. We did not show it what to do. It figured it out alone and found the solution to the problem. With a dog it takes months of patient work before an animal will do what one wants it to do.

On another occasion the team experimented with a mirror. When a grouper was shown a mirror, the fish mistook the reflection for a rival, charged the mirror, and smashed it. But when an octopus saw its mirrored reflection, it first stared at it, then extended an arm and laid the arm flat over the mirror. It then "began a slow, wiping motion across the mirror's surface, like a windshield wiper, as though to wipe off the reflection." When nothing changed, it paused, looked again, began wiping again, and finally stopped, seemed to reflect, appeared distressed, and then returned to its hole and refused to come out.

Another researcher, M. J. Wells, has demonstrated that an octopus is capable of distinguishing the size of objects and of choosing between complicated forms composed of lines and designs even when in a stooped or vertical position. Andrew Packard recorded that one octopus was able to distinguish between a vertical rectangle and a horizontal rectangle, and that it did not forget this piece of learning over several weeks. P. H. Schiller has reported octopuses capable of moving down a corridor, looking for a passage, and making detours to reach a crab they had seen through glass.

Other experiments have shown that blind octopuses, by their sense of touch alone, could perceive whether or not objects could be moved by them. Moreover, the senses of taste and smell, as well as the chemico-tactile sense, are very important in their perception and awareness of their environment, although for humans, who are but poorly endowed with comparable senses, it is difficult to interpret correctly the animals' responses when they are mediated by these.

Finally, there is the feeling that humans have of cooperative friendliness when they work with octopuses. Katharina Mangold, who has spent twenty years breeding and observing octopuses at the Banyuls-sur-mer (France) laboratory, says, "They know me very well. There is no denying that a relationship develops between the octopus and the observer.... But it is very difficult to interpret the responses observed ... for the simple reason that we have even less affinity with octopuses than we do with mammals."

Jacques Yves Costeau has commented on this that it is true that "cephalopods live in another world. I mean not only that they live in the sea, which is now open to exploration by human beings, but also that they inhabit a world of sensations and perceptions that is not our own." But he added, "The evolutionary path which led cephalopods to a high degree of perfection is not that taken by the human race. It is nevertheless parallel to ours; and it may lead them further still."

He also wrote,

> If, in spite of its gifts, its highly developed nervous cells, its ability to remember and to judge, the octopus is not the ruler of the sea, it is perhaps because their blood is of a composition that does not carry oxygen as effectively as does the iron in human hemoglobin, so that the larger species tire easily. Also, and even more pertinent, the life span of an octopus is only about three years. . . . It is fascinating to speculate on what the octopus could be if it lived longer. Its increased experience might then make it the sage of the sea.

Perhaps this is an opportunity for us to think again about all the other forms of intelligence and other worlds we have projected so far. In the beginning, we were inclined to pass over the octopus form as a basis for a higher intelligence; now we feel that this judgment was hasty. An anthropomorphic bias is powerful in viewing an alternative intelligence; even with a conscious endeavor not to be so, one is almost inevitably influenced by it.

In its way an octopus is a living example of what we are trying to do, in our readers' company, in this chapter: that is, putting together, like a jigsaw puzzle, pieces of life's manifestations as we know them, and trying to visualize them rearranged in other patterns. Octopuses, it is true, have arms and suckers that are different from claws, paws, or hands, but their efficiency is as great as that of hands; they have beaks like birds, venom like snakes, chromatophores like some fishes, eyes like mammals; they walk and swim, and they are gifted with intelligence. But they are invertebrates. They are totally unlike any other form of intelligent life familiar to us. The senses of touch and sight are all we have in common.

And yet octopuses already possess many of the characteristics that were features of mankind's own march to culture. Their bodies are vulnerable and so they are impelled to construct shelters; they have useful prehensile limbs, and in an elementary way they use tools. We have the feeling that, given long ages of further evolutionary development in which, perhaps, their life spans became lengthened and their energy span bolstered, their resulting increased experience might show them ways to compensate for their bodies' lack of protective armor.

They are, in any case, playful and curious. It is quite possible that eventually they might turn their propensity for scavenging and utilizing whatever comes their way into a more sophisticated use of tools, and in the long run discover an ability to modify some of these objects to aid them in fashioning artifacts according to their needs. None of these possibilities stretches our imagination too far, and were they to occur, we should have all the bases for the eventual development of a culture and a civilization.

We have mentioned previously that one of the necessities for a high intelligence is a prolonged and protected childhood. In the short, average three-year span of their lives, octopuses as we now know them lack this. But we do see that they show great protective concern for their eggs. Female octopuses do not lay them and abandon them, but attach them to protected places, stay near them, fasting to death while they guard them until they hatch.

At their present stage, octopuses' vulnerability imposes a shy solitariness upon their natural lives, but in their contact with humans they demonstrate a capacity for friendly social interaction. Contact senses in higher creatures are associated with warm sociability—the kind of family life led by such of earth's whiskered, burrowing, and intelligent creatures as the badger, which lives with its mate and offspring in a cleanly maintained nest and frequently sorties with its young to pay visits to neighboring badger families. By extension we could visualize a high culture for our intelligent tactile beings in which sociability played an important part.

We come to this conclusion because tactile interactions are of such importance in the lives of so many creatures, including mankind, that they are taken for granted and not remarked. We notice only the dire results that occur when they are missing.

Since our present subject is the receiving and sending of information by tactile means, we shall not go into the social importance of touching in detail, but we cannot speak of tactile communication without mentioning the great significance of mutual grooming in the social order of primates, and the desolate state to which those are reduced who are deprived of it; the "licking into life" of many mammal newborn by their mothers; and, not least, the enormous importance of the part played by the loving handling of human infants in their healthy emotional development.

The licking, touching, and handling of infant mammalians immediately after birth, and in primates, the continuation of interpersonal tactile contacts throughout infancy, seem to be the mechanism by which the new creature's peripheral nervous system is stimulated into an ability to respond to its environment as a separate being, and thus to become aware of its surroundings and of itself as an entity. Thus the tactile sense is aroused by tactile means, and this paves the way for all the other senses and the emotions they arouse to develop in normal sequence and to full potential later on.

An awareness of the self—perhaps in this context we should call it the sense of the self—has an important role in the development of intelligence. We have no way of knowing at what stage in the evolutionary process a sense of self becomes manifest. We know that a mouse or an octopus has an intelligence that permits it to cooperate in an experiment and that an octopus can learn to recognize individual humans and to play, but does a mouse know it is a mouse, or an octopus, that it is an octopus? This we have no way of telling.

We do know that when a chimpanzee sees itself in a mirror for the first time, it investigates all around the mirror and does eventually come to the conclusion that it sees itself. In experiments chimpanzees have been seen unmistakably to recognize themselves. Some animals, notably dogs and otters, when they are brought up from birth by humans, do seem to identify themselves with the human who raises them and handles and touches them repeatedly, stroking, patting, or slapping. Such animals compete with their human handlers' own children for attention, want to sit on their laps or in their chairs, sleep on their beds, even eat their type of food—food that would not be appropriate in a feral state.

Of course, this kind of behavior is due in large part to these animals' innate intelligence, but the tactile contact with the human from birth modifies the creature's sense of itself as a being. In human babies, as in nonhuman primate infants, an absence of touching from birth leads to a crippled emotional development, and in humans to autism.

In this chapter we are writing of a variety of existing senses and thinking of them as pieces of a jigsaw puzzle that may be interlocked in various ways to compose pictures in our minds of other possible kinds of life in the universe. The tactile sense is not a dramatic one and not the first we notice among these pieces, but it is a very important one and should not be overlooked.

HEAT SENSES

When it comes to radiant energy, of themselves there are no such things as "hot" or "cold." These constitute a continuum that ranges from absolute zero to unknown upper limits. Our own experience of heat or cold is influenced by our body temperature, and this is true of all other warm-blooded animals. In general, we can say that any substance that exceeds our own body temperature is experienced as hot, and equally that anything that is below it, feels cold. In mankind, as distinguished from some other species, the heat sense is imprecise. Our bodies' necessities do not call for exact determination of temperature, so the sense has not evolved any degree of refinement.

To be more precise, we, and presumably other warm-blooded animals, do not respond with only one sense to temperature fluctuations. We possess two distinct systems of receptors, one for warmth and one for cold. If a sample area of skin is explored point by point with a metal contactor cooled well below the body temperature, a pattern of responsive "spots" is found. Some points will evoke strong sensations of coldness, while others produce only moderate, or cool, feeling.

In 1 square centimeter of skin area some half dozen such cold spots may be found, and in some areas, such as the tip of the nose, the lips, and the forehead, many more (up to 13). If the experiment is repeated with a heated contactor, one or two reports of warmth may appear, but perhaps none.

An explanation for the fact that we possess so many more cold- than heat-sensitive points may be found in the circumstance that we also possess pain points in the skin, which take over when heat (or cold) is too great. Presumably, danger to the human organism is greater from excessive heat than from cold, since pain points take over sooner in that direction. (As a matter of interest, in human beings there are, per 10 square centimeters of skin, 80 cold spots to every 6 hot ones on the forehead, 130 cold to every 10 hot receptors on the tip of the nose, and 190 cold to no hot ones at all on the upper lip. Obviously, excessive heat is so dangerous on the upper lip that in that area only pain points register it.)

Our own bodies' registers for heat and cold may be compared to thermometers that indicate only gross changes. Others of earth's creatures possess thermometers far finer than ours. The mosquito, for instance, can register differences of as little as one five-hundredth of a degree centigrade—an incredibly fine grade of distinction—at a distance of 1 centimeter. Some fish, notably the sole, respond to temperature changes in the water of as little as $0.03°C$. With this ability to register changes in water temperature so exactly, they are able to recognize the changing seasons and adjust their timetables for migration. Another well-known example of remarkable sensitivity to heat is the bedbug. Crawling along the ceiling of a bedroom it unerringly senses a sleeper's one spot of exposed skin and dives directly onto it.

An equally remarkable faculty that is absent in humans is the Absolute Temperature sense. For many of earth's creatures different degrees of temperature have an absolute quality; they are perceived as distinctly from each other as we are able to distinguish the colors of the spectrum. To bring this home to our minds we have to imagine that for some creatures the colors of the rainbow might appear simply as varying degrees of light: such animals would not be able, as we do, to separate any area of the spectrum sufficiently from the rest to be able to give it a name—red, yellow, blue.

If we speak of heat, we can see that we are as insensitive to degrees of heat as those creatures are to color. Yet not all creatures are so, and for those who perceive heat as precisely as we perceive color, given intelligence and language, there would be a conceptualization, and probably names, for the different areas of the heat continuum. A future human space traveler encountering a civilization based on a form of life possessing an Absolute Temperature sense, would surely be baffled by its concepts, even if he mastered its language.

For such beings, objects of the same form but of different temperatures might have different significance, as for us a red flag is a signal of warning, a white one a signal of submission, and flags of various colors symbolize national sovereignty, Morse code language, maritime conventions, and so on. Differences in temperature might also provide a basis for art forms, as shades of color do for us, not only in painting but also in visual appreciation of our environment and in many deeply emotional ways. In such a world we might walk past an edifice, or any area, and vaguely notice slight and, to us, insignificant differences in the heat of parts of it; but for the being of that civilization such an area might be an object of art that conveyed a rich emotional and aesthetic experience.

In our world many insects, some birds, some rodents, and some fishes possess such a sense. Herbert Heran investigated the heat sense of honeybees, and S. Dijkgraaf made exciting discoveries about this sense in fishes. Among other phenomena he demonstrated that fish could be trained to recognize a specific temperature, say 58°F, within 1 degree of accuracy, irrespective of whether the fish came out of a previously warmer or colder environment.

Other zoologists have demonstrated similar abilities in rodents and bees, but an example that strikes our imagination even more strongly is that of the Australian bush turkey, also known as the incubator bird. It builds huge mounds, 2 to 3 feet high, of whatever vegetable matter is available, in domelike structures over its nest and eggs. The remarkable thing is that the bird maintains the heat at the heart of the mound where the eggs lie at a constant temperature of 33°C (91.4°F), independently of the constantly changing temperature of the Australian bush.

To accomplish this the bird digs air vents or closes them and adds coverings of sand for heat insulation, thinning or thickening them as the circumstances require—not unlike a human engineer, who must get his heat information by means of intricate sensors and who must watch a panel registering their indications so as to maintain a constant temperature for delicate machinery, for chemical processes, or for the heating and air-conditioning of buildings. Of course, in modern times automatic feedback systems such as thermostats duplicate for us what the bird can do by its own recognition and manipulative endeavor without any artificial apparatus.

The Australian zoologist H. J. Frith built three electric heaters into a bush turkey's mound, switching them on and off at random. In spite of this interference, he was unable to change the temperature in the egg chamber faster than the bird could correct it. Every few minutes the bird would investigate the mound by sticking its beak into it in various places and bringing out small sand or soil samples. It seemed to evaluate these samples by testing them over its tongue or the roof of its mouth—much as a human wine taster rolls a mouthful of wine around—and then letting them dribble out. The bird's sensitivity to heat differences was so great that the sensations of its tongue were within a tenth of a degree as accurate as a thermometer, and the bird then acted upon its information as efficiently as if it were a human expert in thermodynamics.

Speculatively, this particular sense presents quite a challenge in terms of imagining the technology, and especially the architecture, of intelligent beings endowed with it and with the ability to regulate external conditions to achieve a specific temperature. Staying strictly within biological parameters, a fictional portrayal of such creatures could open a treasure trove to the fledgling science fiction writer and yet remain eminently realizable.

The perception of precise degrees of heat, albeit in some creatures extraordinarily accurate and in others less so, may be compared, as we have suggested, to a built-in thermometer. We now come to a heat sense that is radically different. It is the heat perception of a snake that operates on the principle of an infrared camera.

The organ through which this perception is monitored is a reflector-shaped membrane in a pit beneath each eye. Each membrane contains 150,000 nerve cells sensitive to heat in an area in which a human would have approximately only three heat points. The difference, however, between the reptile's heat sense and the warm-blooded animal's is not only one of degree. It is of a different nature.

Like the camera, it gives an outline image of a creature that may be only a fraction of 1 degree centigrade warmer than its environment. This enables the snake not only to hunt for prey in darkness, but also to unmask the prey animal's protective camouflage in daylight. For an intelligent being possessing this sense, it would be the equivalent of a built-in infrared camera and would give a picture of the world in many ways similar but in some distinctly different from the way we see it.

For example, if we photograph with an infrared camera a bed which the person who has slept in it has left, the outline of the person appears plainly on the picture. Similarly, the intelligent infrared-sensitive individual would have a clear view of the recent past via this kind of sight much like the olfactory creature's image of it via the sense of smell.

Heat persists as smell persists, and refinement of senses of this nature would have to shape the kind of intelligence and awareness of the world of an intelligent being endowed with it. In which way would surely depend upon other factors such as the particular environment, the bodily structure of the beings themselves, and other external factors. In this, we can only let our imaginations take over.

MAGNETIC SENSES

Senses involved with an awareness of magnetism are the hardest of all for us to conceptualize. We human beings have absolutely no way of physically registering the properties of magnetism, nor can we imagine what kind of sensation they might produce in a living organism. All we can visualize is that perhaps it works in a way similar to those fishes that electrify their immediate environment and are sensitive to distortions in the electric field they create. Perhaps magnetism-sensitive creatures have absorbed iron in an organic form, which is concentrated in one of their organs in sufficient strength to generate a low-intensity magnetic field. This thought is not too farfetched, since warm-blooded creatures, including man, possess an organ that is rich in iron—the spleen. The spleen is the main organ in which overaged red blood corpuscles are destroyed, thus releasing the iron that is bound up in hemoglobin. Alternatively, we might visualize this sense as an awareness of a force that by its differential penetration could give an image not unlike an X ray.

However, the magnetic sense we actually see on earth is a little different from this. Here, the several species that possess it appear to be sensitive to earth's magnetic field and to orient themselves by it, rather than to generate their own. It is possible, therefore, to imagine intelligent beings that have this sense as an accessory one, especially if migratory habits put a premium on an ability for orientation.

On earth an awareness of magnetic forces appears to have evolved especially in some migratory creatures; it seems that it is brought into use when other guides or landmarks aiding orientation are obscured. This idea was tested and found valid by the zoologist F. W. Merkel of Frankfurt University.

It was generally known that during the autumn migration period the robins in the vicinity became restless and had a tendency to direct themselves southwestward, which is the way they would have to fly to reach Spain from the Frankfurt area. But when the robins were placed in a steel chamber, largely impervious to magnetism, although their migratory restlessness remained unchanged, they flew off in all directions so long as they were inside the chamber. As soon as the steel door was opened they resumed their predominantly southwesterly flights. Thus it was clear that whatever force it was that guided them was impeded by the steel barrier. While it is true that it could be some force unknown to us, it is much more likely to be earth's magnetism, which is known to be blocked by steel.

Subsequent experiments proved that these birds responded not to gross distortions of the magnetic field (as when magnetic bars were placed around their necks), but to subtle changes. For example, the Frankfurt robins responded to a terrestrial magnetic force of 0.41 gauss. Inside the steel chamber this force dropped to 0.14 gauss. After several days during which the robins were disoriented, they gradually resumed their generally southwesterly movements, apparently adjusting to respond to the far weaker force, much as the human eye accommodates itself after a while to darkness.

It seems that a magnetic sense is also present in some mammals. Weddell seals submerge under Antarctic pack ice and unerringly swim over considerable distances to ice-free locations on Antarctic coasts. J. Bovet trained wild mice to follow specific compass directions when all other senses were blocked out.

While on the subject of the ability to respond to physical forces that aid orientation, perhaps we should mention organs of equilibrium that are sensitive to a planet's gravity. Mammalians' sense of equilibrium is centered in the labyrinthine organ of the inner ear, and other creatures possess equivalent organs in other anatomical areas. However, the essential function of all of these is merely to inform an organism which way is up and which is down. Obviously, this is an important sense, but it is not one that gives information about the features of an environment. It must be present, but it cannot contribute to the development of intelligence.

GROUP INTELLIGENCE

Intelligence, which includes ways of perceiving, of understanding, and of doing, is a function of complexity. In its broadest sense, of course, the term "intelligence" can be used at every level, from the simplest awareness of a single stimulus like light, through every degree in between, to the incredible complexity of the human brain. But our knowledge of the course of the evolutionary development of intelligence on earth shows us that at each stage along the line, as a new factor is added, the progress is not one of simple addition; rather, radically new capacities become possible.

In grossly simplified terms, a single neuron of any brain is capable of the performance only of stereotyped functions. A thousand neurons do not perform merely at a thousandfold increase of the single neuron's capacities, but at an exponentially different level, because the multitude of interactions between them come into play along with the simple addition of their numbers. Thus a brain of, say, 800 cubic centimeters' capacity is a brain of a different order from one of 400 cubic centimeters; it does not merely perform the same functions twice as well.

The same is true for the brain or the equivalent nervous structure of any individual creature. Complexity, however, may arise from an aggregation of comparatively simple individual units as well as from an increase in the components of a single one. If we take, for instance, a creature with a relatively uncomplicated nervous structure, what it can achieve on its own is limited; but if such a creature is integrated into a large population of its kind—a population within which it is possible for separate groups to be formed, each with a specialized function—then an existence on a totally different level may emerge. One may think of this as the outcome of a "group intelligence" or a "mass brain." In all biological systems the whole is inevitably exponentially greater than the sum of its parts.

If, with our individual-oriented point of view, this seems like an alien concept, we have only to look around us to see the principle at work in varying degrees. At its lowest degree it is present in any society, including our own, where because of the varying skills of its members, the society as a whole produces works that no single individual within it could produce alone. Without the skills of an architect, a mason or a plumber, an electrician or a carpenter could each manage to erect a building on his own; but with the architect to plan the overall design, to coordinate the abilities of all these craftspeople, and to select and supervise the use of building materials produced by a number of manufacturers, the resulting building will be one that no single person could achieve by himself. The same is the case in every other field of endeavor and, of course, in the society as a whole.

In most of mankind's societies, of course, if an invader were to conquer and remove a society's key members, social breakdown would result. But the breakdown would be only temporary because, given time, new individuals would emerge equipped to take over the functions of those who had been removed. However, there have been in the past and perhaps there still are some societies where the element of "group intelligence" operates at the next higher degree.

In the Incan civilization of ancient Peru, for example, as well as among the Mayas, the Zapotecs, and many of the other wonderful civilizations of Central America, the ability to direct the specialized skills of the people was concentrated by heredity and training in the brains of an elite group, a handful of highly educated priest-leaders. The populace, divided into specialized castes, each able to perform only the specific tasks for which its members were destined from birth, was left without capacity to survive as a group whenever the misfortunes of war or natural disaster removed the small elite group.

Civilizations that produced Machu Picchu, Chichén Itzá, and Monte Alban disintegrated and their works were left in ruins or were overtaken by the growth of jungle. It is almost as if the priestly elite operated as the neocortex of the group brain of these societies; when it was cut off, the social groups were left without the ability to originate and coordinate the complex functions they had been capable of performing.

The organization that leveled off the mountaintop at Monte Alban without benefit of powered tools or indeed of any metal tools at all, that planned the temple city on that man-made plateau and supervised the assembling there of the materials, the skills, the arts—the total culture—that went into the construction of its pyramids, temples, observatory, palaces, ball courts, and ritual places—all that disappeared. The total society became like a brain-damaged person who can no longer coordinate complex thoughts but can only survive on a primitive level of function. The difference lies in the fact that the brain-damaged person has suffered an *organic* loss of function, whereas the operation of the elite-led social group is determined by the presence of a culturally cultivated element.

Like the Incas and the Mayas, other societies have also achieved great architectural works with very few tools other than the muscle power of the human body. In one of the oldest and greatest of human cultures, that of ancient Egypt, the dominance of an elite few imposed order and direction upon the physical labor of the many in order to produce with very few tools other than muscles works that still amaze us by their magnitude and technical perfection.

In the tomb of Seti I, where one may still gaze at decorated ceilings and walls, admire complex passages and chambers, and marvel at the wealth and skill of a culture that so long ago could build such sarcophagi and surround them with such art, the excavators saw, still impressed in the sand of the floor, the footprints of the slaves who had carried the mummy to its resting place three thousand years earlier.

Some indication of the sheer magnitude of the labor force

that built those monuments to human power and skill may be gleaned from Herodotus, who recorded an inscription he found on one pyramid, detailing the quantity of radishes, onions, and garlic consumed by the hundred thousand slaves who toiled for twenty years to build it. Accumulated wealth, the result of the success of the society, was also a factor in the power to organize the transport of huge stones weighing many tons over a distance of 600 miles, and to raise some of them to a height of 500 feet.

Another ancient historian, Diodorus Siculus, wrote that "an inscription on the larger pyramid . . . sets forth that on vegetables and purgatives for the workers there were paid out over 1,600 talents"—a sum that in the 1930s was calculated to be the equivalent of sixteen million dollars and that expressed in today's dollars must amount to more than double that sum.

No more than among the Mayas and Zapotecs could these vast enterprises have been achieved by human labor alone. Among the Egyptians as among the Central Americans, a trained elite caste provided the equivalent of the organizing function of an individual neocortex to the total social group.

We shall now take this idea of a "mass brain" one step further, to a stage where evolutionary processes have modified individual bodies into specialized forms, fitting them exclusively for specialized tasks. Individual nervous systems have developed that respond to the pressures of a group's necessities, so as to impel those bodies to perform those, and only those, tasks. The "mass brain" is then organic and totally operative.

We refer, of course, to the extremely complex insect societies where any individual is capable only of stereotyped and limited activities, but where the society as a whole shows an order and a functioning on such a high level that it gives every appearance of being organized by an overall "group intelligence." Such a society survives even if a large number of its key members are removed, because it is regulated organically and the lost individuals are regenerated. This time we shall take the termites as our paradigm.

Probably the most basic necessity of any society of living creatures is that the density of its population be adjusted to the capacity of its habitat to support it, and automatic, self-regulating mechanisms have been evolved to meet this necessity in all animal societies. In societies where the function of individuals is specialized, however, there is a further imperative: that it produce exactly as many soldiers, as many nurses, as many fertile females (queens), and so on, as it needs, and no more.

Human societies are inclined to leave such regulation to the laws of supply and demand. If this fails, they expect their leading individuals to cooperate and, through forms of government, to use their brainpower and organizing abilities to correct any anomalies. The results of this system, as we know, are somewhat haphazard. Among the social insects nature's own methods regulate such matters and leave nothing to choice.

All termite individuals, as they hatch from their eggs, begin life as undifferentiated nymphs, and all of them have the potential to specialize into reproductives, or soldiers, or members of other castes, through further stages of development. In every one of the two thousand known species of termites, each has at least three structurally different forms that live together as castes, with different functions in the colony; some have as many as six or seven castes.

Although each caste possesses its own specialized pattern of behavior (its "instincts") that conform with the eventual bodily shape it takes, yet these caste differences, both of form and behavior, are not hereditary: the specialized development takes place only if and as it is needed.

The number of individuals that actually go through all their potential phases is automatically controlled by the scent language of the pheromones. Like some ant species, termites foster and maintain a special caste of soldiers whose bodies are adapted with weaponry to defend their societies. But it is very important to them that they do not raise too many soldiers—that, so to speak, might overtax their economies—but that nevertheless they raise a number sufficient for adequate defense. This intricate balance is maintained by a refined use of their scent language.

Those that develop into soldiers constantly secrete this scent substance, and when the number of soldiers present is sufficient for the colony's requirements, the soldier odor reaches a certain level of strength that has the effect of inhibiting the growth glands of the developing nymphs of the new generation. Instead of continuing to develop into outsize termites, their bodies equipped with weapons, they remain small and become workers.

Thus, a critical concentration of the pheromone acts like an instruction that commands the shutting down of the growth hormones of the immature individuals, and the recruitment of soldiers, which cannot feed themselves, is held at the level of the colony's overall interests.

However, should the termite army suffer reverses and lose a large number of its fighters, the diminution of the soldier scent causes an equivalent number of the dwarf nymphs that were on their way to becoming workers to continue developing beyond that stage into soldier giants. Other castes of termites have their own scent regulators, and recruitment to their ranks operates in the same way, so that the community is kept adequately supplied with the specialists it needs and the termite state as a whole is adaptively populated.

Every other aspect of termite community life is also ordered, as though by a planning intelligence. For instance, some termites, notably a California species, use sound signals as well as their very flexible scent language. When becoming aware of threatening disturbances from the outside, they knock their heads against the ceilings of their galleries as a warning to others to retreat into the more central chambers of the hill. So that this rapping will not be mistaken for accidental bumps or thuds, they use a code of two or three knocks with timed intervals between them.

One of the most impressive aspects of apparent planning in termite city life is its astonishingly efficient engineering. Professor Martin Lüscher, of Berne, Switzerland, has described a striking analogy between man's central heating systems and the air conditioning of termites' nests, which he investigated with a species of termite, *Macrotermes natalensis,* that he found on the Ivory Coast of Africa.

A medium-sized colony of this breed consists of some two million individuals that need a steady "hothouse" climate of 86°F (30°C) for their well-being. To protect themselves from the fluctuations of the external climate, they utilize the accumulated heat produced by the metabolism and activity of each of these small creatures, and regulate it by means of an incredibly sophisticated ventilation and "air-conditioning" system that they build into their hills.

A termite hill has walls as hard as concrete; there are no apertures in them that could, like the windows and doors of a human habitation, easily be opened or closed to adjust the inside temperature and airflow. Yet the millions of insects that live within them must breathe, and they need something over 260 gallons of fresh air each day to do so.

To obtain this, the termites construct within each of a series of about a dozen ridges, discernible from the outside of their hill, some ten narrow air vents that run the entire height of the hill from top to bottom, just inside the outer surface of its wall. These are connected at the top in a chamber just below the roof and at the bottom in a kind of cellar vault.

Thus, the interior of the hill is ventilated by a system of over a hundred passages built into the outside walls, connected by chambers in the basement and under the roof, which form a perfect air-circulating system. The stale hot air rises and accumulates in the roof chamber, whence it floats into the upper parts of the air vents. There it makes contact with air from the outside atmosphere, which seeps through the infinitesimal pores of the hard wall material, and in the process is cleansed of excessive carbon dioxide and takes in fresh oxygen. These areas have been called the "lungs" of the termite hill.

From the roof chamber, the refreshed air sinks down to the cellar vault, about 3 feet below the ground level. From there it circulates upward again throughout all the chambers of the termite edifice.

Throughout its extent, the system is never left unattended.

Worker termites, like busy human engineers, supervise and correct the ventilation system, alternatively narrowing the vents down into an equivalent of valves, or widening them, or completely closing or opening up vents, according to the time of day or night, to the changing seasons, or to adjust the internal climate to fluctuations in temperature and oxygen levels within the hill.

The extraordinary precision of the termites' endeavors is demonstrated in the fact that the temperature that is ideal for them is maintained without deviation right in the heart of the nest, in the queen's chamber. This amazing, cooperatively attained temperature control highlights one of the extraordinary aspects of the existence of social insects: while as individuals they are cold-blooded animals, the whole colony as an entity has the ability to adjust and maintain its internal climate at a steady level, just as a warm-blooded creature has.

What is the "other way of knowing," by which the termites "know" what they have to do and when they have to do it? Instructions cannot be brought to them quickly enough by messengers, since the distances within the hill are far too great. There is no perceptible means of communication that we can discover.

As we see, the "civilization" of the termites is no mean achievement. Indeed, until the coming of mankind, the monolithic cities the termites erected—some of them over 18 feet in height and with diameters of as much as 100 feet across—represented the greatest modification of the natural landscape wrought by any higher form of animal life. When considered in relation to the size of their builders, they are stupendous. A building of comparable size in relation to the height of a man would have to be approximately 4,000 feet high—the height of a good-size mountain. We humans with all our technology have not achieved such a masterpiece. The termite with no tool but its own naturally modified body has done so.

The operation of a "mass brain" is not solely an alternative means of organizing basic group life and the architectural structures to accommodate that life. Not rarely in nature a group brain functions as an instrument for decision-making in a way startlingly like an intelligent individual brain.

How, for example, does a colony of ants arrive at a decision?

The same naturalist, Charles Hogue, who reported his delight over the luminescent beetles, bacteria, and fungi of the Costa Rican rain forest, also posed this question after many months of observation of the army ants there. He wrote that when an army ant colony makes a coordinated move, it is clear that it has to make a number of decisions. A critical one is: "Where shall we bivouac tonight?" Hogue explained that a colony in a migratory phase must push out from its bivouac site every day and take up residence elsewhere every night. He gave the following sequence as typical: a raiding column is sent out, scouts go out from the bivouac and practice their raiding activities, and also along the route at a certain time of the day a suitable log, overhanging branch, or other protection is presumably found. It has to be explored and evaluated as a bivouac site for the next day or for that very night.

Think of the logic involved in this complicated chain of events. The site must be found and recognized by a member of the colony, or two or many members, and evaluated. Then a decision has to be made. Is the decision made in the colony after messages are sent back from the scouts, or do the scouts make the decision and send it back to the colony? Here are more than a hundred thousand ants, which must all, by the end of the day, embark on the same goal. They must cooperate in the same decision. Who makes the decision, when is it made, and how is it made? Further, when it is made, how is it transmitted and explained to all the members of the colony so that each knows how to act?

Hogue writes, "I think that some sort of statistical assessment is reached by a large number of exchanges of information, and the interpolation of these bits of information with stimuli and signals from the environment, such as the time of day, the amount of light striking the forest floor, the temperature, and perhaps other factors." He suggests that "somehow a decision comes out of the colony almost in the same way that an answer to a complicated mathematical problem comes out of a giant computer through the individual but additive actions of thousands of electrical circuits. . . . Such complexity must also exist in the individual and collective circuitry of the nervous systems of the ants."

Speculations have been made that ants communicate by some kind of antennal language. As far back as 1810, Huber, in his book *The Life and Death of the Indigenous Ants* (printed in Paris and Geneva), promulgated the idea that they "talk" by touching their antennae, and the idea became quite popular. Recent attempts have been made to discredit it, but contemporary field workers believe it is likely that there is indeed some form of antennal communication between ants, and possibly by the actions of the front legs as well.

Ants are known to communicate, although not very commonly, also with sound, by stridulation, in addition to their most important language, the chemical language of pheromones. Nevertheless, none of these would seem to be adequate for the almost instantaneous mutual consultation and decision involved in such unified action by tens and hundreds of thousands of individuals.

The decision cannot be made by the queen—she would have no way of knowing, visualizing, and evaluating a site many yards away from her; nor can one believe that there is a sort of council of ants charged with decision-making for the colony. Neither is the computer model an adequate one. We would suggest, rather, that the decision is not an attribute of anything that exists physically in the repertoire of the ants, but that it is an *emergent quality* of the complexity of the interactions between the multitude of sentient individuals, in many ways like a thought—a product of the complex interaction of the billions of individual neural cells, which does not reside in the properties of any one of them.

All these examples—the Incas, the Mayas, and the Egyptians, as well as the termites, the ants, and the other social insects—represent not other senses but rather other ways of organizing the labor of beings possessed of senses familiar to us.

They bring new thoughts to our minds when we think about what kind of intelligent life we might meet when one day we journey to other solar systems.

The civilization of the ancient Egyptians lasted longer than any other human one. Evidence of its sophistication as early as 4500 B.C. survives and, protected by deserts on all sides but the north, where it was isolated by the Mediterranean Sea, it lasted, albeit with an interregnum between the Old and the New Kingdoms, until the first century of the present era when, softened by the long ages of immunity from attack, it fell to a handful of Roman soldiers.

Had such a civilization arisen on another planet in an absence of warlike neighbors, it might perhaps have persisted for long ages more, and we should see there a fabulous way of life, personal luxury, splendid architecture, and unsurpassed arts, in the almost total absence of technology other than its simplest tools—fire, the wheel, rolling logs, platforms, levers, and some cutting tools.

The ancient Egyptians are lost to us, absorbed by conquest and marriage into the Muslim world of the Arabs. Only their monuments remain to inform us of the wonders they accomplished. Their arts and skills and many of their ideas reached Crete and Greece and were transmuted in those areas into forms that eventually became the pillars of our modern laws and arts. The people themselves, however, disappeared from the face of the earth.

With the termites and the other social insects the story is quite different. The ants, for instance, are at least as successful a species as man in adapting to diverse climates and habitats. Their colonies thrive on hills and plains, deserts and forests, all the way from the Arctic tree line to Tierre del Fuego, the tip of South America, and to Tasmania, and they are to be found on virtually every oceanic island from Iceland and the Aleutians in the north to Tristan da Cunha and Campbell in the south. Numerically they are the most abundant; they contain a greater number of genera and species than all other social insects combined, and their ecological and social adaptations are astonishing. The food specializations include feeding on isopods, arthropod eggs, other ants, seeds, secretions from other insects that they domesticate and "milk," or special fungi they cultivate on insect dung or vegetation—the equivalents of our meat, eggs, grains, milk, and cultivated crops.

The complex societies of the ants evolved early in the history of life and have proved so perfectly adapted and stable that they have persisted basically unchanged for hundreds of millions of years. From this fact alone their perfect efficiency is apparent, and forces us to contemplate the possibility that on other planets, no matter what kind of senses predominate there nor what kind of being uses them, their intelligence may be organized in the same way: that is to say, it emerges organically as a function of the complexity of their group life and is not an attribute of individuals as we are accustomed to think of it.

The kind of creature so organized would not have to look like a termite or an ant. Theoretically, such an individual could be of any shape at all, provided only that its final form were reached via a series of stages in which its body could be modified according to the group's needs and its nervous system not be of a type that could eventually evolve into an individual brain.

If such were the case, of course, should we arrive in such a world we should have no way whatever to communicate with its inhabitants. Each individual would be structurally designed to carry out its own task only, without specific comprehension either of its own activities or of their relevance to the group's. The group's "intelligence" resides not in the minds of "leaders" that direct it, but in the organic functioning of the whole.

RANGE AND INTENSITY

In building up images of other possible civilizations emerging from other kinds of intelligence than ours, we should also keep it in mind that it is possible—even likely—that those senses that are familiar to us may arise again in other worlds but in an extended range or at a higher level of intensity. If that is the case, the sense would not merely be sharper than the one we know, but it would take on a different quality.

We have already given a small indication of these possibilities: bees, for instance, are aware of a different portion of the spectrum from that visible to us, so that to them the "reality" of the world is completely different from our "reality." We have also discussed how the dog's world is a world of odors utterly unlike anything we know. The bee's nervous system, of course, limited by its small size, cannot approach ours in integrative capacity; so as an individual, if not as a group, its capacities are restricted. The dog's limitations lie in its lack of hands. But these examples do give an idea of the "other worlds" revealed by an extension of familiar senses.

If we extend this concept and assume that any and every sense we are aware of probably has a range beyond our present comprehension both in extent and in intensity, we can see opening up before us further uncharted sense-domains for the exercise of our imaginations. Sight, for instance, raised to a greater power, might include not only keener sight but also the ability to see other things—things that are too small to be seen by our eyes, things that travel too fast, or things that are too transparent or too opaque to be registered by our senses.

Nor should we forget the disjunctive vision of the deep-sea fishes we have already described, where the brain pieces together information from characteristic patterns of color and light, and "reality" is inferred from an almost Morse code–like abstraction; or infrared vision, by which one could physically see the past. All these produce *visual* ideas of the world, as well as our own photographic type of sight, but are nevertheless totally different not only in method but also in the resulting perception. And the variety and range that are possible for sight are also possible for every other sense.

Before concluding this chapter and leaving the reader with all these "pieces" to sort out, we mustn't forget to emphasize that no matter how exotic, how highly specialized, or how wonderful are some of the senses we have described, any and all of them are ultimately derived from the responsive potentials of the simplest living cell.

The most primitive single-celled organism responds to such basic stimuli as light, touch, or heat. Out of that aboriginal property of the living cell, all other complex senses and sense organs eventually develop. As we have mentioned, the nature of the environment is a major determinant in the type of sense that becomes prominent in any organism.

We should also stress that no advanced, intelligent creature possessed of one sense alone will achieve that state of complexity. A single sense, no matter how highly refined, would be insufficient to bring all the raw information necessary to feed the mental apparatus with sufficient data for intelligent discrimination. Of necessity, complementary senses, also highly developed, must support, confirm, modify, and duplicate the major one.

We must never underestimate the importance of redundance in the evolution of intelligence. We shall be returning to this idea later, but here we stop to recognize that we "know" an object not only by the way it looks, but also by the way it feels, smells, and sometimes to some extent by our emotional response to it.

Imagining another world is a complex game. It is also an extremely challenging one. Moreover, as we suggested at the beginning of this chapter, one or some of our ideas might turn out to be fact. In any case, the exercise is a useful one for a species that has set its sights on the exploration of other corners of the universe.

EIGHT

Other
Ways of Telling

WHETHER WE ARE scientists or lay people, when we think of the possibility of contact with intelligent creatures from another world, one of our first thoughts is: How shall we understand their language? How will they understand ours?

It goes without saying that the modality of expression does not have to be, and rarely is, the same as the modalities of perception. While we human beings perceive the world via sight, hearing, smell, sound, and so on, we communicate with the world *vocally*. We use many kinds of nonverbal communication also, as do all animals, but our deliberate or voluntary expression is via spoken language.

For this reason—and perhaps also because so many other species also use various kinds of vocal expression—we come to think of it as almost a prerequisite for communication. We usually find an unexpressed assumption that another intelligent being, no matter how totally different its form may be from ours, will express itself vocally, as we do.

This assumption has no merit. Human beings came to the ability to use speech and language because a previously four-legged creature began to run and then to walk on two of its legs only and eventually assumed a completely upright posture. This fortuitous development, causing a rearrangement of the structure of the vocal organs in relationship to each other, became the means by which the production of far more, and more complex, sounds became physically possible. Without the change of posture no change in mental development, of itself, could have produced the complexities of human speech.

Why should we suppose that this happenstance, coinciding as it did with other circumstances that favored brain development (the use of the freed hands and eventual increase in brain volume) to produce the highly refined vocal communication of speech, should also occur in another intelligent organism, formed by an entirely different set of circumstances? In our own world the means of communication are varied beyond belief. Each creature gains from its environment the information it needs and has a means of utilizing that knowledge and conveying whatever information is necessary to further its own purposes.

Almost every one of the senses through which various creatures receive news of their world is also, with modulation, a possible channel for sending out information—for communication. To offer what is to us an exotic example, a creature that receives information via sense organs that monitor infrared radiation might also be able to send messages by giving off heat in varying degrees, or pulses, allowing for sufficient modulation to be useful for communication. All one needs for a symbolic language similar to our own is a system of from fifteen to thirty distinguishable elements in whatever modality, recognizable by touch, smell, electricity, or any other means, that can be arranged in various permutations, as can the letters of our alphabet.

As rich in possibilities for communication as any of the sense modalities may be, however, we wish to elaborate here on a means of communication that is far simpler, and that is grossly underrated by most of us. It is the language of gesture and posture.

On earth such communication systems are highly evolved in monkeys and apes, and form the basis of their capacity to express motivation in individuals and to facilitate social relationships. Without this ability to express mood, monkeys and apes would not be able to engage in the subtle and complicated social interactivities that are a feature of their adaptation.

The fact that for long years human language was considered to be the sine qua non of intelligence, proved to be a handicap when scientific attempts were first made at interspecies communication, especially with chimpanzees. (Advanced work on interspecies communication has also been undertaken with dolphins.) Chimpanzees do have a vocabulary of vocal expressions, and it was probably this that misled early researchers into thinking it might be possible to teach them the sounds and meanings of human speech.

An attempt was made by one couple to bring up a chimpanzee, Vicky, from infancy, like a human baby in their own home. While considerable affection and much mutual learning were established among them, Vicky's linguistic accomplishment did not go further than the ma-ma and da-da of a human infant, and from this point of view the experiment had to be counted a failure.

Many years and much devotion went into these efforts, and when they proved fruitless a wrong conclusion was drawn. It was assumed that the chimpanzees were after all not as intelligent as some of their behavior had led us to believe them to be, and as the size of their brains would suggest was possible.

Quite recently it was realized that even though chimpanzees are anatomically extremely close to human beings, their vocal structures, including the larynx and palate, are positioned in a way that makes the production of the sounds of human speech impossible—a bit like the Englishman who tries to repeat the sequential consonants of a Slavic language and cannot do so, even though he recognizes them and understands their meaning. The Englishman and the Slav have the same vocal equipment, and the difficulty each has with the sounds of the other's language is a result merely of the training and usage of that equipment. For the chimpanzee, however, with its different vocal structures, the exercise that is difficult between humans of different linguistic origin is physically impossible.

That the inability of chimpanzees to enunciate the sounds of human speech does not necessarily reflect a limitation on their intelligence has only very recently become apparent.

Observation of them in their native habitats over long periods of time has revealed that their vocal repertoire forms only a part—the lesser part—of their systems of communication. The more important and subtler part of their social communication is conveyed by a system of gestures that is wide in range, modifiable to adjust to degrees of circumstances, and regulates a complex social life simpler than, but in several ways similar to, our own.

With this knowledge at their disposal, researchers have now been able to embark on a different and far more fruitful tack in their efforts at communication with chimpanzees. In the process they are realizing that these animals can, and probably do, conceptualize very much along the same lines as human beings do.

Various methods have now been devised to provide alternative symbols that convey words of our own language. Some researchers have taught their chimpanzees to use plastic pieces of different shapes and colors to symbolize words and ideas; they have found that with these "instruments" at their disposal, the animals were able to understand and respond to human language.

Others obtained even more rewarding results by utilizing the chimpanzees' own gestural propensities and teaching them a sign language similar to the one used between deaf-and-dumb humans. With this linguistic tool at their disposal, the chimpanzees have helped us make large strides in facilitating interspecies communication.

Now human technology is being brought into play. Chimpanzees are being taught to use language consoles at which they learn that certain buttons project symbols that have certain meanings; all they have to do is to touch the relevant buttons, and conversations can be carried on between man and animal. The "education" of the chimpanzee is proceeding apace.

Lana is a chimpanzee who is being educated at the Yerkes Primate Research Center in Atlanta, Georgia, by a doctoral student, Timothy V. Gill, and who has learned enough of a modified English to converse with people through a computerized keyboard. She has been in training for two years and is now four years old—the physical equivalent of a human first-grader. It was reported recently that she has gone beyond conversations using only words deliberately taught her, to a stage something like that of a human child who suddenly realizes that all things have names and begins to ask, "What's this called?" and "What's that?"

Lana discovered the concept on her own during a session when Gill was testing her on the naming of a bowl and a metal can by asking, "? what name-of this." The modified English used has been called Yerkish, and a Yerkish conversation is carried on by using an electric keyboard, enabling the chimpanzee, with her dexterous hands, to utilize words that would be impossible for her to pronounce vocally. When keys are pressed, the words are projected in sequence on a display panel above the keyboard, and both Lana and her teacher must read the projected images to get the message. The words she uses are projected as geometric images, which she has learned to read as meaningful words and sentences. So, incidentally, have her teachers.

Lana had learned the names of "bowl" and "can" only a few days earlier. Every time she answered correctly with a phrase such as "bowl name-of this," she had been allowed to take a piece of candy from a box. On that day Gill entered Lana's room with a bowl, a can, a new object, and a box of candy. Lana went to her console and pushed six buttons as follows:

> *Lana:* ? Tim give Lana this can.
> *Tim:* Yes. (He gives her the empty can.)
> *Lana:* (Puts can aside.) ? Tim give Lana this can.
> *Tim:* No can.
> *Lana:* ? Tim give Lana this bowl.
> *Tim:* Yes.
> *Lana:* (Puts bowl aside.) ? Shelley.
> *Tim:* No Shelley. (Shelley, another technician, was not present.)
> *Lana:* ? Tim give Lana this bowl. (Before Tim could

answer, Lana typed out another sentence.) ? Tim give Lana name-of this.

Tim: Box name-of this.

Lana: Yes. ? Tim give Lana this box.

Tim: Yes. (Whereupon Lana ripped it open and took out the candy.)

Later that day when Gill brought a cup into Lana's room, she again asked the name of the new object.

This conversation (as reported in *The Stars and Stripes,* December 6, 1974) clearly shows that an ability for mental conceptualization similar to a human's is present in the apes, but because they lack the anatomy for speech, humans had not previously credited them with it. Now that scientists have realized this and have turned their attention to devising several nonvocal methods of communication, great progress is being made. Dr. Duane Rumbaugh, chairman of the Department of Psychology at Georgia State University, who directs the research with Lana, is delighted with it. "We've obviously always underestimated the intellectual ability of chimpanzees," he commented. "It's now clear we're not pushing Lana at all. She can learn a lot more things if we can only figure out how to teach her."

Lana has now become another of the chimpanzee pioneers, along with Washoe, who learned to use sign language at the University of Nevada, Sarah, who conversed by manipulating shaped pieces of plastic, and others, who are helping humans to find a bridge of communication between species.

For our present theme they are also vastly interesting. They teach us to keep in mind that if, in space as well as on earth, we find creatures unable to converse with us vocally, we should not necessarily infer an inability to conceptualize, to think, or to translate thought into action.

A young Japanese macaque began to wash her sweet potatoes and eventually taught this cultural acquisition to all the members of her troop. Subsequently, she discovered how to clean the sand off her grains of wheat by sifting them through water, accustoming the monkeys who copied her to play in the water until a cultural revolution took place (the monkeys began to swim and to pass on this skill to others). We may now speculate on what would be the result if some of the educated chimpanzees were returned to their native habitats and were able to pass on their new language skills to others. The sign language, especially, appears to be so admirably suited to their physical abilities and natural habit of communication by gesture, that it no longer seems at all farfetched to think of educated apes carrying on conversations among themselves as well as with human beings.

What kind of evolutionary trends might this initiate? Perhaps over generations a total gestural language, as opposed to signal sounds and signs, might evolve. It might also be illuminating to think about how much information not only monkeys and apes but also we humans convey by nonverbal means. Beyond the spontaneously invented signs and gestures we use to serve to make our meaning understood when we meet individuals with whom we share no spoken language, there are also our own normal gestures that amplify and substitute for our words.

"Body language" conveys a wealth of meaning via posture, gait, facial expressions, degrees of animation, heightened or reduced skin color, laughter, tears, foot stamping, fist clenching, cowering, table banging, and so on—and on. There are also conventional gestures of greeting and farewell, of triumph, of agreement or negation, all of which are almost universally understood, no matter which language is one's own particular means of vocal expression.

Whatever our speculations about language, or any other kind of communication, however, we have always to remind ourselves that just as perception is selective, so also is the cognitive process. What a creature will make of the circumstances it perceives and what it will communicate will be relevant to the totality of its mode of life, just as its sense perceptions are adjusted to registering those aspects of its environment that are important to its existence or, in the case of humans, that seem to be important in the circumstances.

It is perfectly apparent that we do not at all times see equally clearly all the things around us that theoretically are visible. Our brain selects out from the environment those parts of it that seem significant, just as the cocktail-party goer's ear selects out of the general hubbub those voices its owner wishes to hear. It is that selection that in turn determines the material that reaches the cognitive processes for, so to speak, "digestion," and that in the end determines the *content* of the communication any being will make.

Even within our own species—and even if, for experimental purposes, a large number of people from different parts of the world were placed into the same setting—human beings of one culture will observe, evaluate, act upon, and communicate about aspects of it other than those of another. Language often gives a clue to the cognitive processes of individuals of a particular culture. In English, as in most European languages, for instance, there is but one word for "snow." An Eskimo would be surprised by our insensitivity to the different kinds or degrees of this phenomenon; his languages have many words for it, each of which describes a specific kind of snow. Obviously, snow is an aspect of the environment that is more vital to an Eskimo than to a European; his senses are therefore attuned to closer observation of its finer differences, and he has words to express these.

Not only in the spoken word but also in the written, mankind reveals cultural variations in the repertoire of a human being's conceptualization of the world. The characters of Chinese writing, for instance, represent ideas rather than sounds. A curve over a straight line (the sun over the horizon) means "morning"; the sun and moon together means "light"; a mouth and a bird together means "singing"; a woman beneath a roof means "peace."

These idea-characters have been refined over the ages and the pictorial element has been overlaid with additions designed to define the term specifically, so that today many of the characters are highly complex symbols. Thus there is a sign that represents a horse, another for "a bay horse with a white belly," and yet another for "a horse with a white blaze on its forehead."

Nevertheless, many of the characters remain relatively simple and have the great advantage that the written language can be understood and read by Koreans and Japanese as easily as by Chinese, even though the symbols are read as different words in different localities. It also provides a standard means of communication for all the inhabitants of China itself, where the multitude of dialects are for the most part mutually unintelligible.

As Will Durant pointed out in his monumental *The Story of Civilization,*

> this advantage applies in time as well as in space; since the written language has remained essentially the same while the spoken language has diverged from it into a hundred dialects, the literature of China, written for two thousand years in these characters, can be read today by any literate Chinese, though we cannot tell how the ancient writers pronounced the words, or spoke the ideas, which the signs represent.

Indeed, among objects exhumed at Honan were writings, tentatively dated back to the Shan Dynasty (1766–1123 B.C.) and thus approximately 3,500 years old, in characters substantially like those still in use in our own time and still readily comprehensible over such long ages.

To quote Durant again,

> This system of writing was in every sense a high intellectual achievement: it classified the whole world—of objects, activities, and qualities—under a few hundred root or "radical" signs, combined with these signs some fifteen hundred distinguishing marks, and made them represent, in their completed forms, all the ideas used in literature and life. We must not be too sure that our own diverse modes of writing down our thoughts are superior to this apparently primitive form. . . . Such a sign language unites a hundred generations and a quarter of the earth's inhabitants

in spite of their natural diversity and their variations in space and time, with a means of expressing their ideas in a mutually intelligible way.

In our context, there is a further significance to this type of written language. It is another surface expression of a different cognitive appreciation of the surrounding world. It reflected the profound conservatism and unrivaled continuity of the Chinese civilization, and at the same time contributed to it. The minds of the people who wrote it developed an unsurpassed degree of sensitive perception in the arts, but did not put value on science or industry. They "preferred [says Durant] the quiet and mannerly rule of tradition and scholarship to the exciting and disturbing growth of science and plutocracy."

Such verbal and linguistic distinctions as these two examples reflect, indicate underlying cognitive differences among members of our own species. But they are merely the proverbial tip of the iceberg when compared with what we must expect to find as we begin to explore interspecies communication. Even if we manage to find a language that we can use as a bridge to communication with another of our planet's species, such as the apes, that communication is bound to reveal deep gulfs between us and them when it comes to perception and to the understanding and use they make of whatever is perceived.

The apes are closely akin to us. They and we descend from common ancestors; our bodily structures are similar; their mental processes and ours have developed from common origins. Yet there is a chasm between their cognitive modality and ours. What we shall find when we meet a totally alien intelligence, born of a totally alien type of life on a planet unlike our own, will of necessity be even more radically removed from our own thought and communication processes than anything we can readily imagine in advance. And this is a probability we must bear in mind lest we think too glibly of the possibilities of communication with other intelligent beings of other worlds in the course of space travel.

NINE

Other
Civilizations

THE SPECTACULAR STRIDES made by our own highly technological civilization over the past hundred years seduces us into believing that scientific ingenuity is the hallmark of advanced civilizations. We have come to think that the progressive accumulation of knowledge over generations must lead to ever more wonderful apparatus, and that as we gain more information about life's processes, we shall be able to devise ever more efficient means to control them. We believe that all higher civilizations must tend toward these capacities and these goals, and we imagine that the high civilizations of other worlds will have taken at least as long if not longer strides along similar paths.

The history of human civilizations, however, does not support this idea. It is the delusion of all peoples in all times to believe that the type of society in which they happen to live —its values, its achievements, and its aims—is the best or the most advanced of any that have arisen until that time, and they can only imagine the future in terms of their own society's values.

If we were by a quirk of fate to be transported backward in time into a community of ancient Egypt, we should surely think of future progress in terms of bigger and better river barges, more magnificent palaces, temples, tombs, and monuments, ever greater domestic luxury, deeper understanding of the ways of the gods and of the movement of the heavenly bodies across the firmament—and perhaps of the possibility of obtaining an inexhaustible supply of slaves to assist us in attaining these goals. Were we to find ourselves part of a Phoenician community at the turn of the current era, we should be likely to imagine the future in terms of possibilities for faster ships and more distant voyages. Presumably if in such milieus we thought at all about life on other planets, our speculations would follow similar lines.

If we speak of other civilizations, of course, we should define our terms. For some the word refers to eras when the high development of arts, of skills, and of the luxuries of living was achieved by certain groups of people in certain places over certain periods of time. For others a civilization refers to a more general concept: the sum total of the way of life of any group of people bound together by common customs and laws in a given area over that period of time during which that body of custom and law serves to hold them together in an identifiable group.

But however we choose to define our subject, when we look over the records of the past on earth with the bird's-eye view of an historian, it surprises us, given our own view of "progress" as associated with science and technology, that of all the great civilizations that have been achieved, very few have been technologically based, even though the people of those societies were as dexterous and ingenious as we, and no less intelligent. We have but to see the relics of their works and to read their writings to be convinced of this.

We tell ourselves that those ancient civilizations did not accomplish the technical wonders of our own because they had an insufficient accumulation of information at their disposal. Yet in some instances we see the invention of techniques equivalent to some of ours that were put aside and not developed—not because the ability to do so was not present, but because the ideals and aims of that society put no value upon their development. Where mankind has desired to achieve technical objectives, those objectives have been obtained with or without a sophisticated technology. Given an intelligent being, that intelligence will not automatically concern itself with technology; it will be concerned with accomplishing the aims upon which its society sets values.

The supreme example of the ability to invent almost any of the artifacts rediscovered later in Western societies, and the almost casual neglect of such inventions and indifference to putting them to use, of course, is the early and extraordinarily advanced civilization of China. Paper and ink, block printing and then movable type, books, paper money, dictionaries and encyclopedias, as well as gunpowder, were all first devised and made in China, long before the Western world rediscovered ways to produce them, and attest to the ingenuity and skill of that people. Yet the unique civilization of China was pervaded from its earliest days by ideas of "government by virtue," of the achievement of harmony between man and nature, by emphasis on philosophy, art, and learning. Those inventions that did not further these ideals were neglected as unimportant or superfluous, mere curiosities or interesting "toys."

The spans of civilizations, like the lives of individuals, are transitory. We see today signs of decay in our own civilization and have no clear view of what new ways of life and systems of values will succeed it. It is quite possible that the veneration of technology will fade and be replaced by new visions and new ambitions sooner than we believe. If we look back over the course of human history, we see that some of the longest-lived and many of the highest, by any standards, have been the least technological. Let us take a brief look at some of the other ways of life that have guided the energies and molded the thinking of human beings in other times in other parts of our own world.

The Sumerian civilization was based on an irrigation system that channeled the waters of the Tigris and the Euphrates as early as six thousand years ago. There, abundant crops of corn, barley, spelt, dates, and many vegetables were raised. Plows drawn by oxen and already furnished with a tubular seed drill aided those people to till their soil. They threshed their harvested grains by drawing them over great sledges of wood to which they attached flint teeth to separate the straw and release the grain.

Among the Sumerians metals were rare, a luxury. They had

some copper and tin and occasionally mixed these into bronze. Sometimes they made some large implements from iron. But most Sumerian tools were of flint and other easily obtained and worked materials like clay, ivory, and bone. Yet on the basis of such simple materials and tools, the earliest human civilization of which we have knowledge emerged. Clay tablet records inform us of the accessions, victories, lives, and works of their kings, who at various times from 4500 B.C. onward brought much of western Asia under their control.

We read from their cuneiform writings, made by impressing a wedge-pointed stylus upon soft clay, of their priests and kings, of their edicts, taxes, and laws. We learn of such monarchs as Urukagina of Lagash who "gave liberty to his people," who forbade exploitation of the poor, and who instituted the oldest and fairest, if briefest, code of laws in history; of Gudea, devoted to religion, literature, and good works, who built temples, fostered the study of antiquities(!), curbed the power of the strong, and showed mercy to the weak; and also of bloody conquerors and of majestic rulers like Sargon, whose portrait was found on a monolith at Susa, as well as other great kings, like Ur-engur, who proclaimed, "By the laws of righteousness of Shamash, forever I established justice."

We find the existence of poetry apparently 4,800 years old, and we learn of a weaving industry organized on a large scale and supervised by royally appointed overseers, of goods carried largely by water but also overland to destinations as far away as Egypt and India. Contracts were confirmed in writing, duly witnessed; a system of credit existed by which goods, gold, and silver could be borrowed at agreed rates of interest; gold and silver vessels, ornaments, and weapons were fashioned; society was stratified into many classes and gradations; medicine, although still bound up with theology, flourished, with specific remedies for specific illnesses prescribed; a calendar was devised; property rights were sacred.

A system of law, the wellspring of the famous code of Hammurabi, governed commercial as well as sexual relations and regulated loans and contracts, all trade, all adoptions and bequests. Courts of justice existed, over which professional judges as well as priests presided. Litigation was avoided where possible by submitting every case first to a public arbiter, whose duty was to attempt to establish an amicable settlement without resorting to the courts.

The Sumerians worshiped many gods, but their priests transmitted education as well as theology: boys and girls were instructed in writing and arithmetic, patriotism and piety, and were prepared for professions. School tablets survive and still carry the tables of multiplication and division, square and cube roots, and exercises in geometry that were part of the curricula of schoolchildren in that remote time.

Marriage was regulated by many laws: the bride kept control of her dowry; she alone determined its bequest; she shared with her husband equal rights over their children; in the absence of the husband and adult son she administered the family property as well as the home; she could engage in business independently of her husband and could keep or dispose of her own slaves. Sometimes, as in the case of Shub-ad, who ruled her city luxuriously and imperiously, a woman could rise to the status of queen.

Cosmetics and jewelry were used. Professor Sir Charles Leonard Woolley found in Queen Shub-ad's grave a little vanity case of blue-green malachite with golden pins and knobs of lapis lazuli. It contained a tiny spoon for scooping up rouge, a metal stick for training cuticles, a tweezer for shaping the royal eyebrows. The queen's rings were of gold wire, one inset with lapis lazuli; her necklace was fashioned of fluted lapis and gold.

The richer citizens built palaces atop mounds sometimes forty feet above the plain. Since stone was scarce, they were built of brick, the walls decorated with rich designs, the inner walls plastered and painted, around shady central courts. Water was obtained from wells. Furniture was tastefully constructed; some beds were inlaid with metal or ivory, some armchairs embellished with feet carved like lions' claws. (The poor built their houses of adobe over a framework of reeds—a technique not too different from that used in southern Morocco today. These homes had fewer amenities, perhaps, than a modern country cottage, but were certainly pleasanter than a tenement dwelling in a modern city slum.) Crude vessels of pottery and delicately wrought ones of alabaster and gold have been found in the earliest graves at Ur, some as old as 4000 B.C., along with decorated daggers in jeweled sheaths, seals made of precious metal or stone and with reliefs beautifully carved on their small surfaces.

By 2700 B.C. Sumerians had established huge libraries; at one (at Tello), a collection of over thirty thousand tablets was found, stacked in a logical arrangement. By 2300 B.C. Sumerian historians began chronicling their past and their present.

But these artifacts, portraying for us something of the luxury and comfort of this nontechnological civilization, were as nothing against the sophistication of their writing, capable of conveying still to us their usages in commerce and law, their ideas in poetry and religion, and a record of their government and daily life. With the simplest means—a triangular wedge mounted on a stem to hold it—without inks, paper, or parchment, those early people devised a method of recording the complexities of human thought and of preserving over the millennia knowledge of those first states and empires of which we are aware.

If we have dwelt at some length upon this most ancient of civilizations, it is because we believe it offers us a challenging thought in our efforts at constructing mental pictures of the civilizations of other planets. We, who are still overimpressed with the marvels of our elaborate machinery and electronics, our sophisticated industries and sciences, are forced to ponder whether indeed these modern wonders represent progress or merely change.

Life in Sumeria six thousand years ago would seem to have offered privilege and servitude, luxury and simplicity, scholarship, art, philosophy, crafts, and labor, enlightened or oppressive rulers, judges, priests, farmers, and artisans in about the same proportions and with similar satisfactions or hardships to the citizenry as our modern societies offer us. We have no

Other Civilizations

guarantee that advanced technology is a necessary or even a useful appurtenance of higher civilizations.

This doubt is even more firmly established when we take a look once again at the achievements of ancient Egypt, the heir to the culture developed by the Sumerians. There, in a time more ancient to classical Greece than is the Greece of Sophocles to us, the civilization of Egypt, as Will Durant wrote, "flowered into a civilization specifically and uniquely its own; one of the richest and greatest, one of the most powerful and yet one of the most graceful, cultures in history. By its side Sumeria was but a crude beginning; not even Greece or Rome would surpass it."

Above all else, the quality of ancient Egyptian architecture astounds the mind as it pervades the spirit. That those magnificent temples and courts, majestic porticos and obelisks, the very forest of colonnades and statues, entablatures and bas-reliefs that still survive at Karnak and Luxor could have been constructed as they were in the times of Queen Hatshepsut and her successor Thutmose III, sixteen and fifteen centuries before the time of Christ, with the minimal mechanical aids that were then at mankind's disposal, shatters our ideas about "progress." Long centuries before Doric columns were raised in Greece, they were anticipated here. When Champollion, in the wake of Napoleon's conquest of Egypt, discovered Karnak, he exclaimed, "There all the magnificence of the Pharaohs appeared to me, all that men have imagined and executed on the grandest scale. . . . No people, ancient or modern, has conceived the art of architecture on a scale so sublime, so great, so grandiose. . . . They conceived like men a hundred feet high."

Among the earliest individuals known to history was Imhotep, in every way a forerunner of Renaissance man. He was an artist, an architect, and a royal adviser (to King Zoser, *circa* 3150 B.C.). He did so much for medicine that he was later venerated as a god of knowledge, and at the same time he founded the school of architecture that provided the next dynasty with its master builders. Under his administration stone houses were first built, and it was he who planned the oldest pyramid. In his time richly colored glazed earthenware was produced, rivaling the faience (decorated earthenware) of medieval Italy. Such artifacts would not have been made unless the standards of luxury of the domestic lives of at least some of the people created a demand for it.

Fifteen hundred years later—yet still fourteen hundred years B.C.—in the time of King Amenhotep III, his capital, Thebes, as described by Adolf Erman in *Life in Ancient Egypt,*

> surpassed in magnificence all ancient and modern [cities], her imposing palaces received tribute from an endless chain of vassal states, her markets were filled with the goods of the world, her temples were "enriched all over with gold" and adorned with every art, spacious villas and costly chateaux, as well as shaded promenades and artificial lakes, provided a setting for sumptuous displays of fashion that anticipated Imperial Rome.

Ancient Egyptians mined copper, iron, and gold as far away as Arabia and Nubia. Their metallurgists fused copper and tin to make bronze; their armorers fashioned that bronze first into weapons, helmets, and shields, and their smiths later made tools of it—wheels, rollers, levers, pulleys, windlasses, wedges, lathes, screws, drills capable of boring the hardest diorite stone, saws that cut the massive slabs of rock and marble. Their workers made brick, cement, plaster; they glazed pottery, blew glass, and infused these with color; their carpenters were masterly at wood carving and also made boats and carriages, furniture and coffins so handsome they still arouse our admiration today. Their tanners made leather clothing, seats, quivers, and shields; their artisans made ropes, mats, sandals, and paper from the fibers of the papyrus plant. Other craftsmen developed the arts of enameling and varnishing, applied chemistry to industry, and, four thousand years ago, wove some of the finest textiles ever produced—linen so fine that it requires a magnifying glass to distinguish it from silk.

Egyptian engineering was superior to anything known to

the Greeks or the Romans or, for that matter, to Europe before the industrial revolution; only our own time has excelled it. Under Senusret III (2099–2061 B.C.) a wall 27 miles long was built to drain the waters of the Faiyûm basin into Lake Moeris and thus reclaim 25,000 acres of marshland for cultivation. Great canals were constructed, some from the Nile to the Red Sea; the caisson was used; obelisks weighing a thousand tons were transported over great distances. All this was done without sophisticated machinery.

Courts of law regulated public order; a regular postal service existed. Watercraft of all sizes and degrees of luxury plied the river, from the poor man's simple boat to the great vessels, rowed by dozens of rowers, that were used for the transport of obelisks or the pomp of the rulers. In the streets the rich were carried in sedan chairs or driven in chariots, the pharaohs in carriages of silver and gold.

The personal toilette of the pharaohs and of those of high rank required the services of launderers, bleachers, guardians of the wardrobe, barbers, hairdressers, manicurists, perfumers, cosmeticians, wigmakers. Those who could afford it had put into their tombs when they died seven creams and two kinds of rouge, kohl for the eyes, color for the lips and nails, oil for the hair and limbs, toilet sets, mirrors, razors, hair curlers, hairpins, combs, cosmetic boxes, dishes and spoons made of ivory, alabaster, wood, or bronze. Clothing was made fragrant with incense and myrrh. Men as well as women of all classes decorated neck, breast, arms, wrists, ankles, and ears with jewelry.

Learning was fostered and schools existed for the children of the well-to-do. Paper and ink were available to the students of the higher grades, and sheets of paper were gummed together and bound into books. Eventually, an alphabet of twenty-four consonants emerged from earlier hieroglyphic signs, and passed with Egyptian and Phoenician trade to all parts of the Mediterranean. Via Greece and Rome they came down to us—probably our most precious heritage.

The Egyptians themselves in their writing mingled pictographs, hieroglyphs, and syllabic signs to the very end of their civilization, but without their alphabetic characters (which appear first in inscriptions left by them in mines in Sinai and which have been variously dated at 2500 and 1500 B.C.), our own civilization is unthinkable. Literature flourished; libraries of papyri rolled and packed in jars, labeled and ranged on shelves, have been found. One of the oldest stories is an ancestor of the Sinbad the Sailor or Robinson Crusoe tale, and the earliest form of the Cinderella story, complete with exquisite foot, lost slipper, and royal marriage, was already written down. Short stories abounded, as did tales of ghosts and miracles, romance of princes and princesses, fables of animals, and many songs of love, hymns, and other religious works, tales of the gods, records of the great deeds of the kings and queens, historical narratives, and a great literature of poetry.

We are impressed with the striking similarities between the personal lives of those ancient people and the daily lives of the citizens of any later civilization, including our own. As we learn of the ways in which the early Egyptians adorned themselves, of the occupations and pastimes we see illustrated in their murals, of the literature that interested and entertained them, we have to remind ourselves almost forcibly that these people lived from six thousand to two thousand years ago and that theirs was almost the earliest high civilization and therefore had little if any precedent upon which to model itself.

It would seem that human beings, endowed with intelligence as we all are, find similar ways to occupy their time and to interest and amuse themselves, regardless of place or period and, remarkably, of the development of technology or the lack of it. An intelligent brain, like any other organ, must be kept occupied to remain healthy, and if there is insufficient occupation for it in just staying alive, then it will, and must, invent occupations and preoccupations, pastimes and puzzles, speculations and investigations, projects and play, to keep itself busy and therefore in good working order. When we find intelligent life in other worlds, we shall almost surely find the same or very similar patterns.

In the sciences we find mathematics highly developed early

in Egyptian history. The erratic rise and recession of the Nile, upon which all life there depended, necessitated careful records and continual remeasuring of the land and redefining of boundaries by surveyors. It would appear that this measuring of the land was the origin of *geometry*. From this beginning, mathematicians more than four thousand years ago learned to measure circles and cubes as well as square areas, and also the cubic content of spheres and cylinders. They arrived at a valuation of π (pi) as 3.16, very close to the 3.1416 to which we have refined this measurement four thousand years later.

In astronomy the Egyptians seem to have made less progress than their contemporaries in Babylon, but they knew enough to be able to predict the day on which the Nile would rise and to orient their temples toward that part of the horizon where the sun would appear on the morning of the summer solstice. They distinguished between the planets and the fixed stars and kept continuous records of the movements of the planets over thousands of years. They noted in their catalogs stars of the fifth magnitude (almost invisible to the eye). From these observations they worked out a calendar—another of ancient Egypt's basic contributions to humanity.

The finest achievement of Egyptian science was medicine. Although their comprehension of the functions of the bodily organs was, to us, fanciful, they described the larger bones and viscera accurately and they understood the function of the heart. Their great physicians and surgeons specialized in obstetrics or gynecology, gastric disorders or ophthalmology, leaving to the general practitioners only the care of the poor.

In a papyrus some 15 feet long, written about 1600 B.C. but summarizing much earlier work, forty-eight cases in clinical surgery, from cranial fractures to injuries of the spine, are detailed. Each case is given in logical order under the headings: "provisional diagnosis," "examination," "semiology," "diagnosis," "prognosis," "treatment," and definitions of the terms used. The writer makes note that control of the lower limbs is localized in the "brain"—a word that is used for the first time in literature in this document.

Egyptian doctors had a vast pharmacopoeia at their disposal (the Ebers papyrus lists seven hundred remedies for as many disturbances, and the tomb of a queen of the Eleventh Dynasty contains a medicine chest stocked with vases, drugs and roots, and spoons), but they did not depend upon these remedies alone. They attempted to promote health by public sanitation, circumcision, emetics, enemas, and fasts. Herodotus reports them as "next to the Lybians, the healthiest people in the world."

To write of the art of ancient Egypt in a few paragraphs would be presumptuous. Right up to our own times it has never been surpassed, and is equaled only by that of Greece. Not only were the major arts, architecture, sculpture, and bas-relief, noble in conception and perfect in artistry, but also the minor arts—those devoted to the embellishment of the home, the body, and the graces of life—showed the same refinement of skill and taste. Astonishing luxury is revealed in sumptuous furniture, fabrics, silver and gold work, crystal, vessels of alabaster and of diorite stone ground so fine that light shines through it, and in a profusion of jewelry fashioned from precious stones and metals exquisitely worked. Orchestras and choirs provided music in temples and palaces, and painting reached the highest levels of art. Philosophy was profound and the ideas expressed often hauntingly modern. It could well be that this earliest high civilization was the finest and most glorious of any that has ever arisen on earth.

In the course of interplanetary travel we can expect to find any type of civilization at all, from the biologically built-in potentials for social organization of the social insects to any of the great variety of types of human group life based on cultural factors.

There is a strong possibility that when we do one day find our way to another planet bearing intelligent life, the individuals of that place may live in a cultural ambience of a type where there is not the least interest in space travel or in technological achievements—like the ancient Chinese who knew of the properties of gunpowder but used it, if at all, as an entertainment rather than as a weapon, and who considered all visitors from other cultures as unworthy "barbarians" with whom it would not be of any interest to exchange ideas.

It is especially interesting in this respect that precisely the two longest-lived human civilizations, the Egyptian and the Chinese, were among the least interested in technological advances, even though they had the ability to achieve whatever refinements of life they desired. They concentrated their efforts on quite different things: the Egyptians on luxury in the present life and eternal life beyond the grave, and the Chinese on the refinement of arts and thought as expressed in crafts, painting, writing, poetry, and the refinement of manners, formalizing interpersonal relationships, and similar concerns.

High civilizations, therefore, are not necessarily expansionistic. They may, on the contrary, be xenophobic—or, like the Egyptians, interested in other cultures only insofar as their populations were sufficiently numerous to provide a convenient source for slaves for their own purposes but not to proselytize.

Then again we should consider that on some planets there may be more than one species that has attained high intelligence and that therefore civilizations of totally different types of life coexist in the same world. To imagine this we might try to think of our own world, in which civilizations not only of primates like ourselves existed, but also perhaps of cetaceans, or of some advanced form of the Mustelidae (an order that includes the otters and badgers)—and of what it might be like if there were some mutual interchange between such vastly different civilizations in the same world.

Another alternative might be a world where the highest form of intelligent life uses another highly endowed but not quite so intelligent species to do its work—as if, for instance, instead of concentrating upon building robot machinery to take over our tiresome chores, we studied and made full use of the potential of the higher apes for those purposes. If such had been the case, we should by now have become a civilization of ethologists, biologists, psychologists, and zoologists rather than one primarily of technologists.

Of course, there remains the possibility, and even the probability, that elsewhere in the universe there do indeed exist civilizations technologically as advanced, or even further advanced, than ours. Our sun is one of about 250 billion in our galaxy, and our galaxy is one of billions of galaxies. Among them the number of stars with planets capable of sustaining life has to be vast.

If we take our own planet as a paradigm, however, we see that civilizations, like individual lives or the lives of species, are born or arise but also die out when their vital energies are exhausted. Some give rise to new civilizations, others fade into oblivion. Some have short terms of power and vigor; others, though more rarely, survive for several thousand years. Eventual extinction, however, has been the common fate of all civilizations and appears to be a law of cultural collectives as of every other form of life.

The question we may ask ourselves, then, is: How far along the technological road can we, or creatures like us elsewhere in the universe, actually travel before our own works annihilate us, or before the lines of technological development upon which we are embarked wear themselves out, become effete, and await the renewal of rebirth in other forms?

At the present time our astronomers dream of intergalactic travel, our physicists strive to discover the ultimate particle of matter, our biologists work toward genetic engineering and to produce human beings perfect in mind and body, our medical researchers hope to conquer disease, and our idealists seek paths to permanent peace. It has been the experience of humankind that whatever we are able to dream of, sometime and somewhere some of us are eventually going to be able to do. It may well be that all these dreams will one day be fulfilled on our planet and it might be that they are already accomplished on some other planet. But the likelihood is that there is a natural limit to the span of any cultural development, as there is a natural span for the lives not only of individuals, of societies, and of species, but also of worlds and of the whole universe.

Yet if we look again at this series of deaths, we see in them all not a total extinction but a positive element. Although the lives of individual creatures come to an end, their genes persist in the forms of their offspring that take on new shapes, think new thoughts, and act out new lives. Similarly, civilizations send out fertilizing germs that take root again where the soil is ripe for them. Sumeria faded into oblivion, but much of what Sumeria learned and thought in the course of its history found its way to Egypt and there became the seed of new forms and blossomed into a civilization quite different from the one that gave it birth.

In its time Egypt, too, died a death so complete that no memory of its achievements remained until in the nineteenth century the chances of war brought historians and archaeologists to its land in the wake of Napoleon's army. Generations of scholars since then have unearthed its relics and revealed again its past glory. Yet before expiring, Egypt had passed seeds of its civilization to Crete, where they merged with the native genius and ran the course of a vigorous new existence there; Crete in its turn fertilized Greece, and Greece, through Rome, the Western world. Each newly incarnated civilization was as different from its parent form as Athens was from Thebes, or as New York or London—or modern Athens—are from ancient Athens, and yet each is in large part the product of the one that went before.

If we speculate in terms of technological societies in other worlds that may be following paths similar to ours, a mere extension or continuation of the one we know will not provide the answer. The only certainty we can have is that whatever form such a society may take, it will be transitional. We can have no more idea of what a succeeding civilization on our planet can be like after ours has disappeared than the most sophisticated ancient Egyptian citizen of Luxor or Karnak could have dreamed of the life and accomplishment of modern Tokyo or Paris. Our technologies will be developed as far as they will go and then, quite certainly, another way of life will arise so new and so unforeseeable to us as to be beyond our present imagination.

TEN

The Nature
of Intelligence

INTELLIGENCE AND EVOLUTION

Any exploration of the nature of intelligence is a precarious venture. Before we start out, we must know what we are seeking, and there is no generally accepted definition of intelligence. It is as hard to describe where intelligence begins and of what it consists as it is to define life itself.

There is no consensus about the point in evolutionary development at which life might be said to begin, nor even an exact demarcation between life and nonlife. Did life begin when amino acid molecules first formed self-replicating chains? Where is the boundary between the replication of crystal formations and that of carbon compounds? Indeed, there is no boundary. Life is a flow. It begins in inorganic material and emerges gradually along its course to sentience and intelligence.

The same may be said of intelligence. It begins with the simplest awareness of limited elements of the environment and it proceeds toward awareness of the self and then to abstract ideation by the most gradual stages. We may examine it and analyze it at any point along this continuum, but no one has yet been able to summarize its totality to the satisfaction of anyone else.

Before we go any further we must make it clear that our investigation is not about the qualities of especially brilliant human mental achievements or of the nature of the minds of those gifted individuals we call "bright," "great brains," or "geniuses," but of intelligence as an expression of life itself. By what means does an organism "know" its surroundings and what it must do to exist in them? What is the mechanism by which it makes a decision to take this path rather than that one? What, in fact, is a "thought"?

Thought is nonmaterial. We cannot touch it, see it, or measure it. Yet at its inception it arises from a material organ, from the living tissue of the brain, and as its outcome it directs concrete actions. This is the mystery of the thinking process. It arises in the concrete and ends in the concrete, but is itself intangible.

If we trace the thinking process back to its origins in the brain, we find ourselves back again at the origins of life. Sensitivity to external stimuli is a property of every living cell. Over the ages the forces of evolution fostered the combination of single cells into more complex organisms, and certain cells became specialized for the execution of certain functions within those organisms. The reception of responses to external stimuli became centralized in cells specialized for this purpose—the neurons—at first in extremely simple, and gradually in increasingly complex, nervous systems.

Eventually, as environmental circumstance and the increasing complexity of organisms put a premium on the development of a coordinating center for the nervous system's messages, an area of the nervous system itself began to specialize for this function and over time evolved into the organ we call the brain. The brain, then, is the outcome of sensitivity to stimuli—that most primitive and most basic quality of life itself.

Another fundamental property of life resides in the chemistry of carbon compounds, in their propensity to form chains, to polymerize, and to proliferate. Life manifests itself by profusion, and out of the profusion, or redundancy, comes the inevitability of selection. Germ cells are generated by the millions when one would suffice. Multitudes of seeds and fruits are produced, but only a few find congenial soil and take root. Animals give birth to offspring far in excess of the needs of the species to maintain themselves, and only the best adapted survive.

To suggest that profusion exists in order to create a possibility for selection, of course, is teleological thinking, and in any case is not accurate. The converse is the case. Selection takes place because there is profusion, and this principle holds in many manifestations of life, not least in the brain.

There are ten thousand million cells in the human brain—more by many factors than can possibly be necessary to effect any action necessary for the organism. Since thought arises only as brains become more complex, we have to assume that it is an outcome of just that redundancy. In the same way that millions of germ cells, thousands of seeds, hundreds of offspring allow for the production of a few biological entities or even a single one that will prove viable, so we must assume that it takes millions of brain cells, in some way activated by and operating upon the neural messages they receive, to produce a thought.

We are aware of a thought as a single, often fleeting entity, but it is quite possible that before the thought reaches the level of consciousness its elements are reproduced in the same profusion as germ cells. The germ cell is not a single element; it is a vastly complex assemblage of elements. Nevertheless, in accordance with the methods of living matter, it is duplicated millions of times in order to perpetuate life.

One single neuron will not produce thought; neither will a hundred, nor even a thousand. The phenomenon of thought becomes possible only when a near-infinitude of possible interactions between millions of brain cells and their myriad axons and dendrites (fibers that branch from cells, connecting them with other cells) are present. In some way not yet understood, thought is a quality that emerges from the sheer number of interacting nerve cells, and a single thought can be conceived as their multiple echo.

Yet even this is not the end of the process. Using germ cells as a paradigm, it is probable that even this mass of elements that has produced a thought is itself replicated thousands of times. How else can we explain the phenomenon that when a part of a brain is damaged, or even totally incapacitated, other parts of the brain take over and the thought processes of the organism persist?

Louis Pasteur, for example, in his later years, suffered a massive cerebral hemorrhage that destroyed a large area of his brain, and yet his memory persisted and he did some of his best work subsequently. Also, following psychosurgical intervention, where substantial tracts of the brain are cut, there is no intellectual impairment. Clearly, brain function is but another aspect of the redundancy that prevails throughout nature and appears to be an essential element of life.

Even this conceptualization, however, only describes, but does not explain, the emergence of an intangible thought from a tangible mass of living matter. But although we cannot explain the mechanism by which the quality we call thought arises, we can see the principle concretely in action in the observation of life's forms. At the lowest level, a single-celled organism, such as a paramecium, responds predictably to certain stimuli. At the next stage, slightly more complex creatures evince reflex responses to a wider range of stimuli. At a higher level still, we find nervous systems genetically programmed to respond to environmental conditions by behavior that is increasingly complex, and finally a brain composed of a number of cells sufficient to allow for alternative, rather than stereotyped, responses. It is at this point that it might be said that true intelligence, in the higher animal and human sense, enters the picture, because where alternative solutions become apparent, a choice must be made, and the *ability to choose* between alternatives is at the heart of intelligence in its higher manifestations.

From laboratory experiments made on artificially isolated neurons, it appears that single nerve cells operate on a simple yes-or-no principle. Either they transmit a stimulus or they do not. But it is equally apparent from the behavior of living creatures that as brains increase in volume and in the number and size (arborization) of their cells, the result of the sum of the yes-or-no decisions of each of the cells is no longer a simple majority decision, "yes" or "no." Rather, as the number of yeses and nos increases, sets of combinations and possibilities arise that offer alternatives from which a course of action may be selected; as the number of possible alternatives increases, the ultimate result becomes more and more uncertain, or unpredictable.

We have now taken the step from intelligence to high intelligence, and even to genius. Intelligence selects between alternatives that become apparent. High intelligence or genius works on a basis of so many perceived alternatives that one cannot anticipate or predict which will be chosen.

Until very recently in human history, indeed into our own times, it was thought that intelligence, logic, the capacity to reason, or whatever name was applied to the highest mental function, was the factor that decisively separated mankind from the rest of the animal world. It was held that the power of the mind was an endowment of our species alone, conferred upon us by the Creator, and that with it went special prerogatives: we were lords of creation. Today it is clear to us that mind or intelligence is, like every other attribute of our species, a stage in an ongoing process. We see elements of intelligence emerging in varying degrees in many species of higher animals of all classes. It is not a property of mammalians alone, but is a quality toward which forms of life tend.

Moreover, as living creatures come to depend increasingly on intelligence for their survival, a selective pressure is set up that results in accretions to that part of the brain that promotes it. The use of intelligence, and its physical base in the brain that facilitates it, work on each other in a circular fashion, each enlarging the capacity of the other, over generations. We are heir to all life that preceded us, and we shall pass on our characteristics, both physical and mental, for use and further modification and refinement to all those that succeed us.

The knowledge that neither human beings nor their brains nor any other of their attributes are unique phenomena, but an aspect of life and a part of its flow, is helpful to us in discerning life's overall patterns. We see that these patterns seem to have certain structures, repeated in many of life's most fundamental processes. This could very well be because all life ultimately is based on carbon compounds and so the properties of carbon compounds inevitably influence processes of life. The principles of redundancy and selection and of ultimate constraints are at work in evolution as well as in the brain.

Fantastic as the idea at first sounds, we find the methods of the brain repeating the methods of evolution. The nervous system receives a manifold redundancy of stimuli from the external world, but it is so constructed that it filters out all but a few of them and eventually registers only those the organism can use—in effect, it *selects* from profusion; it retains information that is useful and discards all other information.

The evolutionary process, too, is based on redundant profusion. Thousands and tens of thousands of individuals combine and recombine the possibilities of their genetic endowment until eventually one or some few individuals are produced that possess some slight modification with some—often infinitesimal—advantage over others of their kind. The progeny of these few individuals inevitably become the objects of blind natural selection until their issue, with their new characteristics, permeates and ultimately replaces the population from which it sprang. In this way each tiny step of change has its trial of life and is either rejected and discarded or selected and incorporated into its continuing flow.

So close are these patterns of operation—profusion (redundancy); retaining or discarding (selection)—that we could call intelligence a kind of speeded-up evolution, or evolution a painstaking and slow, but implacable, intelligence.

INTELLIGENCE AS INFORMATION

There is a universality of intelligence, because the nerve tissue of which any brain is composed can receive only encoded signals. No brain perceives a visual image the way a piece of photographic film registers such patterns of light; that visual image must be reduced to electrochemical signals by the messenger nerves in order for them to be able to carry information about it to the brain. Not only a sight, but also a sound, a touch, a smell, or any other information brought by the senses, whether on earth or on any other planet, must therefore be encoded in order to be registered.

Different senses will filter out and relay different pieces of information, but once the information reaches the nerve, it is of no consequence whether the encoding is a result of perception by visual, chemical, electrical, or any other type of senses: the end result is a pattern of engrams in the brain that are then processed.

All these messages that are carried by the nervous system to the brain from both the outer and the inner world of the organism, we have called *information*, because in essence this is their nature. In effect, two major filtering processes take place. The first is effected when the senses select from the myriad available stimuli those bits of information that by their nature they are evolved to perceive, and the second when the still great mass of prefiltered news reaches the cognitive centers of the brain where it must in some fashion be ordered, categorized, meaning derived from it as it relates to the organism, and suitable responses to it initiated if necessary.

This flow of information, of course, rarely becomes incorporated in the cognitive centers of the brain in pure form. As it is processed it is modulated—either dampened or intensified —and also elaborated in terms of such factors as experience, attention, expectation, pleasure, needs, motivation, or the absence of these. The elaborated information becomes subject to evaluation, which to a large extent involves a matching against previous experience, and then, depending upon the interpretation put upon the elaborated information, it is either acted upon or ignored, stored in the memory to add to the organism's sum of experience or obliterated and forgotten.

There is one other possibility available among the brain's processes, and that is the so-called short-term memory. This falls between remembering and forgetting, since it is a method of dealing with information that is useful only for a limited time. We do not have to remember for a lifetime where we put a key yesterday, or to remind the dairyman to deliver the milk, or the content of a German sentence that we must hold in our minds until we come to its verb; so items such as these are stored in what might be called a temporary warehouse only for so long as they are needed, and then turned out in a process of forgetting. Long-term memory, on the other hand, is stored and becomes the permanent library we call experience, which may be referred to throughout a lifetime.

Long-term memory, in fact, is the mainstay that permits a creature to exist in an environment without overburdening its cognitive system by perpetually reexploring and reevaluating every sensory impression. It enables the creature automatically to match incoming impressions against previously experienced ones. New information arouses all an animal's alerting mechanisms and puts it into a state of alarmed readiness for any eventuality, but signals that have been experienced before and matched against the earlier ones need only to be given the degree of attention they warrant. Insofar as long-term memory enables appropriate rather than inappropriate responses to be made, it forms one of the elements of intelligence.

Recent research points toward the great likelihood that long-term memory is encoded by means of changes in the structure of the nucleic acids of the brain cells. If this turns out to be the case, then there is more than just a semantic analogy between the memory of the brain and the memory of the genes. The brain's memory, of course, is ephemeral. Unless committed to cultural recording and passed on in that manner, it dies with the individual, whereas genetic memory, embedded in the germ cells, may be transmitted through rains of generations so long as our planet supports life.

Perhaps we can highlight another aspect of this concept

with an example from familiar fictional figures. In the fabled grasshopper and the ant we have two insects, each possessed of genetic systems equally capable of near-infinite adaptation to circumstances when necessary. The life customarily led by the grasshopper over the whole existence of its species does not make demands on its genetic endowment for the programming of behavior as complex and flexible as the ant's. It does not collaborate in cooperative societies, labor in specialized castes, build tunnels and chambers, go to war, domesticate aphids, cultivate vegetable foods, or perform any other of the elaborate tasks of integrated social life that are customary for the ant.

Yet the genes of the grasshopper do not differ in their methods from those of the ant; its environment has not called out the full potential of its capacity to direct so complicated an existence. It is within the realm of possibility that if we were given the time and the knowledge, with some genetic engineering we might eventually produce a grasshopper that could establish societies in a similar way. The ant, on the other hand, would appear to have realized the utmost potential of the genetic information it is possible for so small a creature to use.

Similarly, as we now know, nonhuman primates also possess a large untapped potential, not on a genetic but on a mental level, for intellectual achievement that the circumstances of their natural existence have not called forth. Among primates, only our own species is in the process of tapping to the full the possibilities of information stored as memory.

Here again we see a unity in life's processes. Memory and genetic inheritance are each achieved by a system of information storage, probably very similar, in brain or in germ substance, that holds available for recall the blueprints necessary for life's continuity. Thus intelligence is not a late-arriving addition to evolution's repertoire. It does not, like Athena, spring from the head of a metaphorical Zeus suddenly and without biological parentage; on the contrary, it is of the stuff and essence of life. The genetic code itself is a form of intelligence, as is memory, and since life is perpetuated via a code of stored information, intelligence must be considered as an inevitable eventual outcome.

INTELLIGENCE AND THE CELL

What we have been doing in this chapter is to look at a huge subject through several windows, through each of which we get a different view of it, so that ultimately we may obtain a better understanding of its whole nature than we could by seeing it from one aspect alone.

What impresses us is that the more angles from which we view it, the more it becomes apparent that there is an underlying driving force common to them all. We might clarify this in our minds by thinking of extremely diverse pieces of machinery that produce a great variety of end products—let us say a milk-bottling plant, a breakfast-table bread toaster, a streetcar, a municipal lighting system. Although these vastly different apparatuses serve totally different purposes (except insofar as they all serve human convenience), they are all ultimately operated by the same force, electricity; the principles of electrical energy are common to the operation of them all.

We now come to what we might call that "inner knowledge" that is chemically built into each of the cells of a living organism and that allows them to interact with and influence each other in extremely complex ways. This is particularly apparent in the embryo, where each cell seems to "know" when to proliferate and when to stop proliferating; how far to go and in which direction it must migrate to take its place in the body as a specialized unit of one of the body's organs; and also when to start or to stop any of the bodily developments and processes through which it must go on the way to its maturity and in the course of its life.

We think we can best illustrate this facet of life's own "intelligence" by describing something of another one of the cell's many functions—its immune-protective capacity.

No form of life is invulnerable to invading organisms; so for any life form to have survived, it must be possessed of some kind of protective mechanism. Indeed, this is so on many levels: in a plant's defensive weapons such as stings, thorns, or poisons; in an animal's ability to fight off or disguise itself from predators; or in patterns of behavior that promote survival more subtly.

It may well have been that during the early emergence of life, at the period when single-celled organisms began to evolve into multicellular ones, symbiotic relationships existed between the simplest forms and were factors in promoting organic complexity. Over time, perhaps due to mutations or to purely chemical factors, some of these symbiotic relationships may have become converted into parasite/host relationships in which the needs of the parasite became detrimental to those of the host.

Today we cannot be certain of the exact origin of the invasion and infection of some organisms by such others as viruses or bacteria so as to impair the viability of the host. But we do know that organisms have evolved certain cell structures that exercise almost incredibly sophisticated defenses against such invasions. It is to these that we refer when we speak about the immune-protective system.

In the most advanced higher animals we find the immune mechanism specialized into two types. One originates in the thymus gland, which manufactures the so-called T lymphocytes. The other type is produced in the bone marrow and the spleen and is referred to as B lymphocytes. These two types of cells at first appear to be identical under microscopic examination, but radioactive tagging reveals different surface markings.

The T cells respond to fungi, viruses, tumors, and foreign tissues. Their metabolic activities enable them to interact directly with these invaders and to prevent their proliferation in a way that at the current time is still under intense investigation. This is the *cellular* immune response.

The processes of the B lymphocytes are better understood. They respond to bacterial infections, viral reinfections, and to

allergens. When any of these invade the body, the B lymphocytes convert themselves into plasma cells that produce antibodies, which are secreted into and carried by the bloodstream
in the gamma globulin fraction of the serum. This is referred to
as the *humoral* immune response.

After a B cell completes its final differentiation and becomes a plasma cell, it manufactures about two thousand
identical antibodies per second until it perishes (within three to
four days). Each B cell is confined to the production of antibodies of a single class. In a later stage of the interaction
between the invader and the host organism, uncommitted
lymphocytes are stimulated by the persisting presence of the
antigen to form additional B cells, called "memory cells."
These enable an organism that has been exposed to a specific
invader once, to respond more promptly and vigorously in the
event of later encounters.

Within the entire immune system each cell is able to "recognize" a particular antigen. (An antigen is a substance that
invades a host organism and has a characteristic protein that
unmistakably announces its identity and is in effect its "calling
card." There are literally millions of such antigens and there are
equally as many immunologically active cells that are capable
of recognizing and interacting with them.)

Thus, among the operations of the cell itself, we find in the
immune system a close analogy with the memory of the brain
and that of the genetic structure. Each system receives literally
millions of bits of information from which it is capable of
selecting what is relevant to its own function, of encoding that
information it selects, and of storing the encoded selection in its
own form of memory. Moreover, the brain, the genes, and the
cell are able to call on those stores of memory and to use them
when the occasion or necessity arises.

The fundament underlying all of these manifestations of
"intelligence" lies in the chemical nature of the carbon compounds, the building blocks and source of life. The life force
itself derives from their propensity to replicate and to form
long chains. No matter how few the number of molecules

comprising an original carbon chain, that chain proliferates (unless it is inhibited) and forms a mass, whether of the millions of body cells in all their eventual diversity or, within that diversity, of specialized brain cells among which millions register stimuli and produce thought, of specialized gene cells that register the organism's template and reproduce it, or of specialized immune-system cells that register a "memory" of antigens and produce the antibodies that will combat them.

We come again to the principle of redundancy, and in recording it, we ourselves are obliged to repeat: the life process is an intelligence process. Wherever we find any kind of life, no matter how primitive, intelligence, constantly refined by life's own processes, sooner or later is bound to emerge, as we understand it, in the highly distilled forms of thought, reason, and judgment.

ELEVEN

Beyond Human Intelligence

SINCE LIFE ITSELF IS an aspect, or a manifestation, of intelligence, wherever we find life in the universe we shall inevitably find intelligence in a stage at some point along its line of evolution. As life arises from inanimate matter there, as well as in our own world, its processes must inexorably be interwoven into patterns imposed by the nature of its constituent material, the carbon compounds, as they proliferate and tend toward greater complexity. This same pattern that weaves itself into the fabric of life is at the same time the pattern that produces and is the structure of intelligence.

Looking at life on earth from our present vantage point, we see a vast continuum. Life starts small, and in its simplest form the intelligence it houses, or of which it is an expression, is only apparent in the minimal responses of that speck of living matter to such elementary stimuli as light or heat.

The single living cell "learns." The repetition of a certain experience inclines it to respond more predictably to that experience—in other words, it becomes more intelligent—and as it does so, it influences modifications of the physical body that houses that "learning." Over eons, by the simple and automatic process of selection, interactions between the environment and the sentient cell foster, in a cyclical fashion, the refinement of organs that sense the stimuli and at the same time the complexity of the body that responds to them.

This complexity of body takes myriad shapes. Any and

every form that is possible is at once inevitable. Any assemblage of cells that is viable has its trial in life, and those that have the slightest edge in efficiency survive to continue the onward process toward greater intelligence, while the rest live their span and either die out or persist as "living fossils"—way stages on life's journey.

Eventually the evolving and continually increasing complexity an assemblage of cells—an organism—takes on in response to the stimuli of its environment, reaches the limit of the possibilities offered by that form. Then, like the less successful organisms that preceded it, it, too, remains with them in stagnation or dies out. When this occurs we find the progress of intelligence shifted to another line, usually less specialized and more malleable than the earlier "highest form."

It is as though, at these junctures, intelligence took several steps backward to find a new base from which to obtain a springboard to leap forward again to greater advances. When the first amphibians left the seas, for example, more complex, better environmentally adapted creatures, the fishes, remained behind, but life and intelligence had accomplished almost all that was possible within the boundaries of those forms and in that environment. The onward process shifted and became housed in a contemporarily less well adapted unit of life, but one that offered greater possibility for further adjustment— further refinement of responses to more and different stimuli —and more complex bodies and bodily organs both to receive the more numerous stimuli and to respond to them in more versatile ways.

Life's main highway on earth shifted in turn from the amphibians to the reptiles, which reached the ultimate possibilities open to cold-blooded creatures in the great and vastly long age of the dinosaurs, that period of some 150 million years, when life's intelligence was honed through their bodies' responses. When they, in their turn, reached the limits of their form, they sank into stagnation and eventual extinction. Then life and intelligence took off again from a line at that time more backward than theirs, but one that offered a more malleable susceptibility to change, this time via the mammals.

We—mankind—are the current end of the mammalian line in cerebral complexity and the development of intelligence. We may well wonder whether life's still more complex forms and intelligences will proceed from us and our descendants, or whether we constitute another "dead end" of a line, so that inevitable and implacable further developments and refinements are forced to find a new vessel to promote their onward process.

In saying this it sounds as though "life" and "intelligence" were tangible entities that had to accommodate themselves somewhere, whereas, of course, they are *qualities*. Yet these qualities in some ways possess the properties of such a tangible substance as water, which inevitably finds and flows along the channels that are available to it. This equating of qualities, like life and intelligence, with a substance, like water, probably arises because, in effect, they are all indeed but expressions of the potentials of life's materials.

The present vessel of intelligence's ongoing process, the human brain, being composed of living tissue, cannot develop beyond a certain capacity without outstripping the body in which it resides. While the brain governs the body, the body at the same time must nourish the brain—each is part of a whole, and neither can have independent existence. The simple need for nourishment by a blood supply manufactured in other organs of the body sets ultimate limits for the brain.

Of course, one way to circumvent this limitation would be for bodies to grow larger and larger to accommodate the increasing growth of a brain. A larger body would not only afford it better housing and provision; simultaneously, it would provide a greater number of meaningful stimuli to provoke and necessitate its further growth, each part acting on the other and back on itself in life's normal circles of cause and effect.

When we suggest this as a possible direction for future evolution, we have to remind ourselves that our knowledge of the way our brains produce thoughts is extremely rudimentary. This is one of the least understood fields of biology. Evidence suggests that an increased mass of brain tissue permits redundancy, which is the paramount element in higher intelligence. On the other hand, nature itself shows us that there are ways by which the need for a large individual mass of brain tissue can be circumvented. The social insects are a living demonstration that a sheer number of cooperating tiny brains also gives the necessary redundancy and permits mass actions that appear intelligent.

Then again, nature has shown us that an individual creature's evolutionary increase in bodily size alone does not necessarily involve an equal, or equivalent, growth in the proportion of that being's brain tissue. The supreme examples of this fact were the dinosaurs. No organ is stimulated to further growth and development unless a need for its enhancement arises that impels natural selection for that further growth. Apparently the bodily size of the giant dinosaurs was sufficient for their survival in their time, and so no selective pressure for further brain development arose among them.

It has been hypothesized that among them, indeed, a second brain center evolved in a separate mass along the pathways of the nervous systems of some of their species. Fossil remains indicate that stegosaurus had a very small cerebral cavity and also a rather large broadening of the spinal canal in the lumbar region. H.E. Kaiser quotes Branca as theorizing that this may have been a second brain. Kaiser himself is inclined to agree. He believes that the extremely large bodies of the saurians "could not possibly have been supplied effectively by the minuscule brain and therefore one must assume that a decentralization of nerve tissues must have taken place." He suggests that since the hypophyseal pouch—an organ of the stegosaurus body—is quite large, it may well have taken over some of the functions of the brain by regulating peripheral organs and perhaps hormonal production. But even if this hypothesis is correct, apparently the modification was either too little or too late to govern adaptively and to keep those mountainous bodies viable; in the end, it did not save them.

There is, of course, still another alternative to the development of a larger body to provide support and sustenance for a larger brain: the development of different proportions between them. If evolutionary processes were to select for a head, say, one-fifth rather than one-sixth of the total body size, then the body, albeit necessarily sturdier, would have some leeway in accommodating a larger brain, since it would have proportionately less of its own tissue to sustain.

To some extent we human beings are both witnesses and exemplars of such a process. Very recently we have come to understand the mechanism by which it has come about.

It is a condition that has been given several names, but which is most frequently referred to as *neoteny*. We ourselves have termed it *infantilization*, since it is in effect a rejuvenation of an evolutionary line arrived at by means of a *slowing down* of individual development until the adult form of a species eventually comes to retain features that were juvenile and passing phases in ancestral adults. The process is effected by natural selection for rate genes that impose a slower rate on an individual's development.

In our own species, we experience a slowing down and prolonging of all the phases of our lives when we compare them with the lives of other primates: a longer intrauterine phase; a longer childhood; a latency period and longer adolescence; and, finally, a longer adult life and a postmature period that does not occur in other creatures. The effect of this prolongation is to delay and eventually eliminate characteristics that were specializations of the adult forms of our forerunners, and to retain a new adult form that remains comparatively unspecialized and adaptable.

Thus, human beings no longer develop the hairy pelt, the large teeth, the heavy bones, the oblique posture, and many other features of the adult ape, but remain throughout life in what, for primates (which includes us) is essentially a juvenile form. In our behavior, too, we retain characteristics that are associated only with the youthful stage of other primates—curiosity, exploratory predilections, an ability to learn.

A by-product of this process is the relative proportion of the head, and especially the cranium, to the total bodily size. Just as in infants of our own species the cranium is larger in proportion to the rest of the body than it is in adults, so also in adults of Homo sapiens the structures housing the brain are in proportion to the body larger than they are in apes, and therefore permit the growth of a more complex brain over the longer period of development available.

We have no reason to believe that the course leading to our neoteny, or the retention of a youthful form, should have reached its apogee in us. On the contrary, we must assume it to be an ongoing process. But this process, too, has its natural limits. The size of the head at birth is eventually limited by the capacity of the bony structures of the birth canal to accommodate it, and the adaptation of the pelvic structure of the female to accommodate a larger head is limited by the necessities of an ability to walk. This limitation may be (indeed, has been) bypassed by the birth of more immature fetuses, but, again, immaturity at birth cannot go so far as to revert back to the laying of eggs, or every advantage of mammalian adaptation would be lost.

Nevertheless, in spite of ultimate limitations, a progression toward greater neoteny would not have to go very far before a new species emerged among our descendants—a totally new kind of being as different from us not only in anatomy but also in intelligence, as we are different from the apes. Our own brain capacity in cubic centimeters, for instance, is not more than double that of the apes, but its growth in intelligence is exponential.

A further increase of our brain size by even a small proportion, implying as it does an increase not only in the number of brain cells but also in the arborization of each cell and thus its geometrically increased capacities for the complexity of their intercommunication, would be sufficient to bring about an intelligence so much greater than our own that it would amount to another dimension. We have named this new species, the product of this line of our speculations, *Homo neocorticus!*

But to return from speculation to biological realities, we must keep in mind that the continued growth and development of the brain, like the increased size or special development of any other organ, is promoted only out of biological necessities for the survival of an organism in any given environment.

When an organ functions perfectly or a total organism exists in perfect adaptation to its environment, no circumstance can be present that will promote selection for any kind of change. In such conditions change can only be detrimental and selection will eliminate it. Further evolution then occurs only if extraneous circumstances alter the environment so that new development becomes necessary to maintain the adaptation of those creatures that inhabit it.

Paradoxically, then, intelligence arises as a result of life's adaptation to environment, but continues to develop only so long as that adaptation is not perfect; difficulties, and even malfunctions, as long as they are not lethal, eventually spur selection for greater intelligence. The interplay between an environment and the behavior of creatures in it is mediated by their nervous systems, and it is in this interplay that intelligence is finally honed.

The brain itself, of course, cannot do anything. It is a gelatinous mass that needs executive organs. If its executive organs are inadequate to protect the total creature, either the animal will die out or—as has happened in some instances, including our own species—brain development stimulates the use of tools. In the latter case, it is inevitable that the tool use itself restimulates the brain to further development in a continually reinforcing cycle.

The high degree of individual intelligence displayed in human beings may be seen as an essentially compensatory mechanism that has as its function the overcoming of those disadvantages that are inherent in our bodies' lack of specializations. Because we have no tusks or claws, our intelligence is invoked in self-defense and we devise and fabricate weapons. Lacking a pelt or hide, we must find or construct shelter, or die.

Again we detect the circles of cause and effect interacting without beginning or end: a larger brain requires a larger cranium—the need for the larger cranium promotes more immature bodies—more immature bodies require more highly developed intelligence to defend and sustain them—and so we are back at the beginning, again needing a still larger brain to engender that necessary intelligence.

Mankind's intelligence is being brought into full use in all types of cultures, whether in the technologically oriented societies of the Western world or in tribal societies, wherever they still survive. Technology is but one external manifestation of the skills a highly developed brain confers upon its executive organs. The actual biological mechanisms of thinking are no different in an inhabitant of a tropical forest or in a scientist of a Western culture. The tribesman uses as much ingenuity to recognize and follow trails, to hunt, and to make all the decisions necessary to meet the requirements of existence in a so-called simple society as does the engineer, with the accumulated technical knowledge of countless generations at his disposal, to construct the most complicated machines. The external results are different, but the processes of the brain are the same.

We see the most eloquent demonstration of this in the children of the most primitive tribes of, say, Australia or New Guinea, who have been able to make the jump from a Stone Age world into one of familiarity with modern knowledge in one generation. What seems to us to be the long step from the earliest tools fashioned by our forerunners, those chipped stones and flakes that they used as hand axes and cutting instruments, is in fact one continuous line.

Intelligence begins with the sensitivity of a single cell, and by a process of *biological* accumulation and selection, transmitted genetically, reaches a current culmination in the complexity of the human brain. In the same way, the first chipped flint, by a process of *cultural* accumulation of knowledge passed on verbally and subjected to the selection of experience, reaches a current culmination in spacecraft, cyclotrons, and satellite-relayed television. At the same time, the natural and necessary playfulness of the young mammalian as it explores its environment and learns how to live in it, over time, and also by a cultural process of imitation, memory, and the transmission of knowledge and skills from generation to generation, reaches another apotheosis in the high cultures and great arts.

For the comparatively weak creature who must cope with an environment and rise to any situation or perish, the resources of mental equipment are applied first to the most urgent exigencies. The earliest application of intelligence is devoted to devising tools as weapons; those that follow aid the amenities of life; finally, religion, philosophy, and the arts make their appearance.

The kind of tools that are made will depend on the material that happens to be available, be it leaves, branches, stones, clay, or the presence or absence of minerals. Thus an interaction exists between environment and creature that affects the development of tools and culture as well as of senses and intelligence.

In ancient Egypt, for example, the presence of the papyrus plant led to the fabrication of ropes, mats, sandals, and eventually paper; the presence of flax made possible the eventual perfecting of supremely fine linens. In Central America the presence of lava led to cutting tools of obsidian. Once a cultural habit becomes established, however, it remains impossible to predict either the route it will follow or its ultimate outcome. Who could have foretold that the sweet potato–washing habit established among the members of the famous Japanese macaque colony, and the subsequent sifting of sand from grain in the sea, would so accustom these forest animals to playing in the water that they would eventually begin to swim? And in the face of this unlikely outcome of a recently established behavior pattern, who would dare to predict to what the new-found ability to swim might lead?

Yet strangely enough—or perhaps not so surprisingly, in

view of the basic uniformity of mankind's cerebral mechanisms—no matter where, geographically, nor when, over the entire span of human history, local cultures have developed into advanced civilizations, we find that mankind's greatest thoughts, as epitomized in the writings of philosophers, show remarkable similarity. Some views on the nature and value of knowledge, intellect, and intelligence propounded by wise men of China, so distant from us in place and time, are stunning in their modernity and still current validity.

In the misty beginnings of China's long cultural history, Lao-tze, expounding his concept of Tao, the Way (of nature and of wise living), is reported to have insisted that knowledge is not virtue and neither is it wisdom, for nothing is so far from a sage as an "intellectual." The worst government would be one of philosophers, he is said to have averred, for they botch every natural process with theory; their ability to make speeches and multiply ideas is precisely the sign of their incapacity for action!

At a later period the philosopher Chuang-tze, who lived about 370 B.C., showed a sophistication of thinking that we find difficult to credit to those early times. He wrote that problems are due less to the nature of things than to the limits of our thought; that it is not to be wondered at that the effort of our imprisoned brains to understand the cosmos of which they are such minute particles should end in contradiction.

He spoke of the *limits of intellect;* the attempt to explain the whole in terms of the part has been a gigantic immodesty, forgivable only on the ground of the amusement it has caused, for humor, like philosophy, is a view of the part in terms of the whole, and neither is possible without the other. The intellect, said Chuang-tze, can never avail to understand ultimate things, or any profound thing, such as the growth of a child. In order to understand the Tao, one must "sternly suppress one's knowledge": we have to suppress our theories and *feel fact.* Education is of no help toward such understanding: submission in the flow of nature is all-important.

And Wang Yang-ming, who lived from 1472 to 1528, practically summarized our present thesis when he wrote: "The mind itself is the embodiment of natural law. Is there anything in the universe that exists independent of mind? Is there any law apart from the mind?"

Experts feel certain that the human brain has not undergone any significant biological change since the time of the Neanderthals, who, as evidenced in the excavations at the Shanidar caves high in the mountains of Kurdistan in northern Iraq, built fires, cared for their sick, conducted funeral rites, and put flowers with the bodies of their dead. In the last twenty to thirty thousand years, we know from archaeological findings of both historic and prehistoric periods, a high degree of intellectual accomplishment has been uniformly present in all the branches of our species throughout its existence. The brief excerpts we have given of ancient Chinese thought about thought would certainly seem to bear out this opinion.

There is, however, an outcome of the cumulative nature of culture that we must not overlook. Increasingly, as intelligence adapts a living creature to its environment by the use of artifacts and through the cultural transmission of knowledge, the creature and the environment modify each other at an exponential rate. Perfect adaptation, of the order of the ants' or the termites', which has existed in a balance between being and habitat unchanged over hundreds of millions of years, is not possible for mankind. The rapidity of change in our cultural habitat presents a perpetual and continuing stimulus to bodily and, above all, to mental adaptation, which of necessity increases the demands on the new brain to devise accommodation, and in the process speeds both the rate of change and the need for new changes.

And so, although we recognize that we have been mistaken in thinking that the technological advances of Western man might indicate new departures in the human brain's capacities, yet we have also to recognize that that technology is rapidly creating a totally new *environment* for our species. This new environment may well ultimately affect our species' future development; precisely this technology may prove to be the turning point through which, in negotiating it, we may find ourselves in a process of extinction as Homo sapiens and in a stage of transition toward Homo neocorticus.

There are, of course, far too many imponderables involved to feel confident in predicting the future course of our species. Among these are the course of technology itself and how far from natural processes it can carry us before it becomes subject to its own limitations. There is the matter of human population density and whether it will be adjusted by natural means or can be adapted to the biosphere that is our habitat by cultural or social means. There is the question of the medical preservation of the "unfit" and whether we can remain viable at all as a species with the increasing maladaptive dilution of our gene pools. And there is the possibility that ecological interference may ultimately make mankind's existence untenable.

Our evolutionary development may be reaching the end of a line for biological if not for cultural reasons, but we ourselves are inclined to discount this. We believe subtle biological factors to be operating that are not yet clearly discernible, but which may be recognizable in retrospect. Another factor as simple and probable as the advent of another ice age, for instance, would effectively alter and reequilibrate the balance between man and nature, and must also be kept in mind as a possibility.

We believe that the possibility and even the likelihood remain for a true evolutionary progression in the anatomical and physiological configuration of the brain, much like the progression that occurred between apes and man. In that case the departure would be just as radical, and it would have as a consequence new behavioral response patterns that at this point we cannot visualize and about which we can only speculate.

To assume that this new superintelligence would occupy itself with creating a new, weird, and wonderful technology is a naive exercise of human fantasy. Ultimately technology exists to serve the greater comfort of individuals, and a superintelligence may well find other means of achieving this end. Thus, were such a superintelligence to be found in some other planetary system, we might be confronted with something totally alien to our understanding, and even to our imagination.

It seems to us that we can predict as a next development the extension of an actual and currently observable trend. We see in mankind the persistence of a brain function that has lost its original purpose. This function is seated in the brain's so-called limbic system—the areas of the brain that govern our emotions. This still active part of our brain mediates all those built-in and semiautomatic responses that enhanced an ability to survive in our remote animal past. It achieves those responses by setting up in the animal (and, vestigially, in us) an emotional state that promotes the performance of the necessary actions. It is an inherent property of animal behavior patterns that they are coupled with governing emotions.

As human beings, we tend to regard our own emotions, coupled with our social propensities, as a particularly high and noble facet of our behavior, and it may well be that this is so, but we have to assume that an animal has *far more* specific feelings and passions than are known by humans, since a generalized "mood" must precede any action it takes. Konrad Lorenz has quoted Heinroth as saying, "Animals are emotional people of low intelligence," and he himself observed, "We should assume that the realm of human feelings must exhibit a similar process of simplification and abolition of differentiation to that known to have occurred in human instinctive behavior."

The limbic area of our brain still regulates our emotions, and so it is, in effect, the seat of our "humanity." Because of it we feel parental and sexual feelings, know pleasure or distress, fear or anger, start at strange sounds, dislike unfamiliar sights and smells, experience love and hate and the entire complex of emotions of which human beings are capable. It is perfectly apparent that the limbic system is still very much in operation, and yet we nevertheless find equally apparent a growing tendency for the newest part of our brain, the neocortex, to overtake and usurp many of its functions.

The neocortex is the seat of our reasoning abilities. It mediates our capacity for judgment, for making decisions based on logical thought, for solving problems, for evaluation and choice. In human beings there is an increasing propensity for neocortical reason to impinge upon biologically older limbic emotions in determining our ultimate behavior. Indeed, there is hardly an area of human function that is not now, at least to some extent, subject to neocortical control.

If we feel a surge of anger, we are able, if we will, to stay it and to question whether it is a justifiable response. It goes so far that such basic functions as sex and reproduction may be modified by our reason. Even as we feel a biological attraction to an opposite-sex person, our reason is questioning: Will he or she enjoy the things we enjoy? Will he fit into our social scheme? Will she be as interesting a companion as she is good to look at? Whether or not we follow up that initial attraction with invitation or acceptance of further acquaintanceship is often determined by our answers to those promptings of reason.

To dramatize and highlight the coexistence of the two conflicting bases of contemporary human brain function, we have only to look at our scientific community. On the one hand, those engaged in the so-called pure sciences, like mathematics, physics, or astronomy, rigidly exclude emotional involvement and stress logic in the conclusions they arrive at and in every aspect of their work. On the other hand, practitioners of the psychological sciences, especially those concerned with mental health, place inordinate importance on "freeing" the emotions from the shackles of the reasoning mind, in the belief that the reasoning mind is the culprit and chief cause of mankind's maladjustments.

The dual and often conflicting mental motivations indeed reach as far as the ultimate purpose to which our natural life is geared—the production, sustaining, and rearing of a new generation, which today is also subjected to the processes of reason. We ask ourselves such questions as how many children we can afford to raise and, indeed, whether we wish to have children at all. No longer do our emotions alone mediate even the continuity of our species.

Of course, in the vast diversity of human beings, there are many who still are governed in their actions more by their emotions than by their reason; those who are more neocortically controlled are still a minority among us. There is a whole spectrum between these extremes, but no aspect of the old brain's function except the purely autonomic is not to some extent influenced and in the process of becoming dominated by the new.

The most likely next phase of intelligence in our own species, then, is toward the *total* control of our emotions by the neocortex, that is, by judgment, reason, and logic. Such a process would eventually eliminate entirely such responses as love, hate, maternal feelings, anxiety, and also many of the derivative mental consequences of those emotions, such as elation, joy, enthusiasm, as well as depression, psychosomatic diseases, and so on. With this development, should it materialize, everything that makes us what we are would disappear and our descendants would appear to us to be feelingless automatons. They, on the other hand, from their vantage point in time, would look upon us as primitive forerunners of their species, much as we regard apes and monkeys.

Naturally, in saying that the neocortex would eliminate our basic emotions, this does not imply that all emotion would necessarily be expunged from the repertoire of our posterity. The brain has pleasure centers, and these, if denied the stimulus of the old emotions, would surely find new material in purely neocortical responses. The brain's pleasure centers might then find new function in elaborating such neocortical pleasures as aesthetic appreciation, scientific research and theorizing, the special joys of puzzle solving, and similar satisfactions of the judgmental and reasoning mind.

It could well be that when the neocortex finally takes over

completely and is instrumental in producing our hypothetical new species, they will be better adapted to life than we—freed, as they would be, from mankind's unabating internal dissonance between feelings and reason. The apes are admirably adapted to the circumstances of their lives by the operation of their limbic system and with but little demand upon the resources of the budding neocortex of which they are possessed; this future race will equally experience the benefits of having only one dominant cerebral function—in their case, the reasoning neocortex. The whole course of human history may prove to have been but a comparatively short stage of shifting between these two states of balance.

All other species have no option but to be at ease in the niche of life in which they find themselves: the ape does not question its reason for being or its destiny—it simply lives—and the future race will be unlikely to experience such mixed feelings as we know. In the final count, Homo sapiens may be distinguished not for "brain" and "culture," but as the creature that knew dissatisfaction, that lived in "divine discontent," precisely because, in our line, the emerging neocortex has reached in us a stage where it challenges and to some extent controls the old brain, but does not yet have the field to itself. We are in our minds and in our whole being basically dual, and this, beyond any of our technical achievements, is the mark of Homo sapiens.

Whether such a line of development as this occurs in other parts of the universe, where higher intelligences have evolved, is open to speculation. It could be that it is a development that is idiosyncratic to ourselves, but we are not inclined to postulate unique phenomena, and we think it more likely that a stage similar to ours is inevitable in the development of any very high intelligence anywhere. Transitional phases are unavoidable in any line of evolution, and so we believe we should expect to find stages like ours wherever higher life forms and intelligences emerge.

Any transitional stage, of its nature, is not very stable and therefore probably not very long-lived. For that reason, if for no other, the geological record does not provide extensive evidence of linking forms. Hence, the long search for the transitional forms—the so-called missing link—that bridged the gap between the apes and ourselves, a gap that is only now being filled in by very recent discoveries in Africa of the australopithecines and other linking forms.

If, however, we look back at the whole line of primate development less anthropocentrically, we or our descendants may yet find that the australopithecines were a comparatively stable form when compared with us, Homo sapiens. The future may well reveal that we, along with Neanderthal, Cromagnon, and other comparatively recent forerunners, were each and all of us "trial runs" in nature's progress toward a more stable form to house high intelligence. The very fact that we find in ourselves competing mental systems within one brain should alert us to the probability that we ourselves are a kind of "missing link" between the apes and an eventual, more harmoniously adapted Homo neocorticus.

This, of course, poses tremendous problems in planning to establish interplanetary communications. Perhaps it is only in the transitional and transitory stage peculiar to our species —and at that, only for a short phase in the course of its development—that any feeling of need exists at all to reach out and explore. It is a propensity that has marked our progress. Where other creatures have found circumscribed habitats and have adapted to them, mankind has found challenge in new horizons and constantly searched to find them, whether in papyrus-reed craft, balsa-wood rafts, sailing boats, airplanes, or interplanetary rockets.

Now that we begin to understand the vastness of the universe, we feel lonely and isolated on our speck of a planet in one far corner of it. But for interplanetary communication to eventuate, we probably have to seek the long odds that another world has produced another species that reaches our own stage at about the same time as we, or rather at the time that would allow for the journey's distance to reach them. It is not impossible—indeed, it is probable—that in the vastness of the universe such worlds exist.

And where do *we* go from here? What kind of brain will supersede ours?

The superbrain of Homo neocorticus, as we have said, need not be markedly larger than ours in order to produce a mind of a completely other quality. It must, however, be somewhat bigger, both in the number of its neurons and in the axon and dendrite connections between them—and therefore to a certain extent in absolute mass—to achieve that change.

From the ready-made laboratory we have available on earth, we can see in the progression from an apelike brain to the human one, first, a fuller use of the potentialities of the smaller brain. The australopithecines—the so-called man-apes—for example, retained brain dimensions no greater than those of modern apes, yet they fabricated primitive tools and took the first steps toward a human way of life over long ages, before environmental circumstances put a premium on still greater intelligence. Only then did larger-brained species emerge. Among ourselves, in our own species, we are aware of a notable difference in brain performance all the way from the dull, to the average, to the very bright, and up to the genius, while all these brains are objectively of the same dimensions and potentially all capable of performance equal to that of the genius.

It may be illuminating to look at another organ, or organ system, rather than at the brain, to realize how far one can go in utilizing what already exists before it becomes necessary to increase size for additional performance. The muscles of a 90-pound weakling, for instance, are physically no different at birth from those of a 200-pound weight lifter, and they have the same potential for development. The eventual difference lies in the fact that the weight lifter has exploited his muscle power to its fullest possibilities, while the weakling has failed to do so.

In this respect, the brain is no different from any other organ. Its constant use and exercise fosters better performance at a progressive rate. And so we must visualize circumstances in which the demands made upon human brainpower are such as to foster the full use of the tissue already at our disposal—in fact, a time when what we now think of as "very bright" or "genius" performance becomes "average"—before we have to contemplate any increase in size to facilitate greater powers. If we can imagine a society in which what we now think of as genius becomes dull-to-normal intellectual performance, and try to visualize the geniuses of that society, we then catch a glimpse of the first step toward Homo neocorticus.

We can also get a second glimpse of that distant creature in our contemporary earth-laboratory through another of the existing trends we have mentioned: that the *area* of increased development and growth of brain will surely continue the present tendency of the expansion of neocortical function at the expense of the limbic system. After all, what is our present image of a "civilized" person as opposed to a "wild" one? It is of a person who has *control of his emotions,* one who exercises reason and judgment, and who does not get "carried away" by personal passions. Again, if we can visualize a society in which the behavior of the most highly "civilized" of us has become the norm and in which a new, far more highly civilized elect has begun to make an appearance, we may then fill in a second color in our picture of the future of intelligence.

We may project that the brain stem—that part of our cerebral accoutrement that governs the autonomic functions of the body (such as breathing, circulation, and digestion)—may well take over some of the currently more primitive response patterns of the limbic region, while the neocortex absorbs its higher "emotional" functions. Thus, Homo neocorticus would arrive at a truly dominant neocortex—a condition that is now only partly the case, but that may be seen as en route in Homo sapiens.

Given such a brain, let us take a closer look at it. Its enormous capacity to integrate information must eventually make our conventional languages almost obsolete. Our words and sentence structure would become too cumbersome to express the increased speed of the flow of thoughts and too imprecise to frame the increased refinement of perception.

We do not think that language of our present type would ever become totally obsolete, because we believe it to be a strong element in mother-child attachment and bonding, and therefore a stage through which the future infant would have to pass, but we do foresee it becoming less useful as that child gets older. We can imagine scientists of the future studying their infants' language as a clue to their understanding of the psychological processes of their distant predecessors, much as we study embryology to confirm our ideas about anatomical evolution.

We imagine that in that future time, adult communication will be effected by a new type of "language," one much more condensed than ours, and also more precise—perhaps something along the lines of our present mathematical symbolism, extended and made more versatile. After all, our present brain—that is, the brain of Homo sapiens—evolved to meet the needs of our species over the ages before the time of written language. One of its especially important functions in our development was its ability to "package" knowledge in a form that could be stored, and thus enable our forerunners to pass on the experience of their lifetimes to each rising generation. Today, the sheer quantity of information available to us has become so vast that it is quite impossible for any single mind to encompass all of it in the space of a lifetime, either in its extent or in its precision. We imagine that the Homo neocorticus brain will be able to do this.

This means that their increased brain capacity will be put to work encompassing a body of knowledge that increases geometrically *within* each generation. This will probably not be achieved by the addition of larger storage capacity, for huge increases in storage capacity must eventually become cumbersome and in any case limited by physical factors. Rather, we think, it will be made possible by an increase in the ability to digest far larger quantities of data simultaneously and to deduce solutions in an instant.

We now have to commit large bodies of information to memory to see their connections and to match them with experience in order to utilize them appropriately, whether in conversation or in social endeavor. Homo neocorticus, on the other hand, will probably be able to *rediscover* over and over again, without effort, all that the combined efforts of countless scientific minds of our type have cumulatively been able to produce over lifetimes. They will not have to learn and commit to memory, say, a theory of relativity, but would see the principle clearly and immediately, and have the means to be able to express it, any and every time it might be needed. In fact, the concept would be as self-evident to them as the fact that a rose is a rose is to us.

All these things we can see to some extent from the ways in which our own brains are tending. But even in this we must be cautious. Apes, for instance, when they are tested in laboratories, show a great reserve of intelligence and ability that they are never called upon to exploit in their natural habitats. They have also shown remarkable adaptability when experimentally transferred by scientists to other natural habitats greatly different from their own.

In one such experiment, observers removed five chimpanzees to a desolate island in a lake in the northwest of Russia—an area that they described as a "cool jungle." Although night temperatures there often dropped as low as 10 degrees above freezing, the observers reported they adapted so well that "they might as well have been back home in Africa." They made cozy shelters out of branches resembling armchairs and hammocks, reported Leonid Firsov, director of the anthropoid center of the Academy of Sciences of USSR, I. P. Pavlov Institute of Physiology in Leningrad. The temperature of their "beds," warmed by their bodies, reached about 97°F. They found half of the 180 varieties of plants edible, and they dined on leaves, seeds, mushrooms, ants, and dragonflies. Even more fascinating, experiments showed that the chimpanzees skill-fully used sticks, branches, and stones as working implements, and that they had a rich memory of association.

An ape could never know which of its own potentialities would prove useful and become exponentially developed in a higher form. Similarly, we ourselves can hardly know what circumstances will bring a more powerful new brain into being, nor what purposes it will eventually serve. We can only be aware of the enormous reserve capacity that resides in us, and speculate about its direction.

We take it as an axiom that an organ seldom develops, and never persists, unless a function is or becomes apparent for which it can be used. At that point we know that an organ will take on any function that it can perform, and that its use will improve its performance. Indeed, this reserve power that seems to lie dormant in some species until environment provokes its potential into actual use, has been called by anthropologists *preadaptation*. In the primate line, it was the dormant potential of the simian brain that permitted the evolution of the human.

Nevertheless, since we like to speculate, we may fairly entertain the thought that perhaps some of the extreme intellectual feats that are now being accomplished by humans may in turn reach their apotheosis in a descendant species. What seems to us now to be a comparatively useless predilection for, say, philosophy, or for so-called pure science, may eventually turn out for them to be the basis of an entirely new way of life, and in that context *they* may well think of it as a preadaptation. Thus, a genius of today may turn out to be an advance guard, or the embodiment of a preadaptation that is even now paving the way to future Homo neocorticus. Among that future species, what we now consider as the highest, and even unnecessary, levels of mental activity will be commonplace.

In this context it is not without interest that in our time, too, there is a cultural practice that certainly must have an effect of reinforcing by sexual selection the trends that we discern as being brought about by natural selection. The practice is universal education and its extension in the increasing numbers of young people who continue their general education through school and on into universities, advanced technical training centers, and other institutions of higher learning.

In these places those of our young people who have shown superior mental abilities are brought together and segregated for several years just at that period of their lives when they are physically and emotionally ready for breeding, so that the natural outcome of this social practice is to foster mating between individuals who to a certain extent have been preselected for their intelligence. Over comparatively few generations this inbreeding of our more gifted young people must have some effect on the mental abilities shown in the succeeding generations of their offspring. Thus, two powerful forces, natural selection (by which the more adaptively endowed survive in greater proportion whenever competition between breeding groups arises) and sexual selection (by which males and females in their choice of mates tend thereby to fix in the opposite sex those traits that they find desirable and attractive), in our species and in our time, give indications of being mutually reinforcing.

Is this as far as we can extrapolate from what we know of Laboratory Earth in our attempts to see into our own future or into other worlds?

Actually, it is not. It is within the realm of our reason and imagination to catch glimpses of still more distant possible descendants.

We see that in nature successful solutions often arise by what is called *convergence*—that is, they arise independently in different lines that are often separated in time and place and are in no way related to each other, because these lines have encountered similar environmental problems and have coped with them in a similar way precisely because that solution is an efficient one. Prehensile tails, for example, have developed in many kinds of monkey (although not in all), in the common opposum, and in several related species, in the true chameleon and a few lesser genera, and in a fish, the sea horse. Similar needs in these disparate creatures have evoked a similar response in the structure and function of the tail.

In earlier pages we have from time to time described what amounts to a group brain as it exists among earth's social insects, which produces intelligent solutions, although the individuals comprising the groups are separately only capable of comparatively stereotyped actions. What if a group brain were to evolve where all the individuals were possessed of the kind of superbrain and superintelligence we are attributing to Homo neocorticus?

Utilizing the concept of convergence, we see that the social insects, by the use of chemical substances—the pheromones—that give off distinctive odors, have achieved almost instant communication, an important element in their cooperative actions. In a mass brain of cooperative superbrains that might evolve from Homo neocorticus, we could imagine an absolutely new type of communication emerging, no longer either verbal or even more refinedly symbolical. It could be a direct brain-to-brain awareness promoted by an individual brain's electric charges that might be intensified or modified so as to constitute an instant transmission without any kind of language, receptor mechanisms being evolved within the brain itself.

In order to conceptualize this thought, we paid a visit to a modern laboratory where electroencephalograms of patients being tested for brain disturbances are being recorded and processed all day long. We saw there a technician in her middle years who had spent all her working life recording and interpreting the "squiggles" registered by inked points onto long rolls of papers as they responded finely to the electric emanations of the patients' brains. The lengths of paper were subsequently folded into bound volumes.

This technician could open any of those records and read them, understanding all they implied, as easily as we read any page printed in our own language, and she was teaching her young assistants to do the same. In other words, the direct electric emanations from half a dozen different areas of the patient's brain simultaneously, with their respective periods of excitement and calm, had clear meaning for her, and she could read like a book any of the volumes in the large library of records of brain activity that were assembled on the shelves of her office.

If we can imagine these inked lines of the EEG squiggles as an extremely primitive and crude form of what might ultimately evolve as a form of direct brain-to-brain communication, we believe we have a plausible model on which to base our speculations about our descendant species-after-next. Such instantaneous communication could easily suffuse an entire group and make possible intelligent mass action of an order that we can no longer imagine, or even speculate about.

The question that remains is: Why such a brain? What possible environmental circumstance could ever evoke it?

And the answer is that once a higher intelligence appears on the natural scene, it makes its own environment. It is a matter of common observation among ourselves that every human being leaves a mark of his or her own personality upon surroundings. A housewife's individuality can almost be read by the condition in which we find her house—the order or disorder in which it is kept, the way it is furnished and decorated, the implements and equipment she uses. A man's fields or garden, office or workshop or den, also reveal much about him.

It would seem that something of the brain and perhaps of the whole nervous system of each of us is displayed in the environments we make for ourselves. Even the little bowerbird reveals his individuality in the small objects he uses to decorate his bower and in his arrangement of them. And so we feel we are justified in assuming that as a brain evolves in complexity, it will surely be to the accompaniment of an increasingly complex ambience.

Human beings originally lived in the same "nature" as all other animals, and many still live very closely with it. If we camp in a forest, we do not actually need our superior mental equipment—that other creatures live out their lives in that forest is evidence of this. But most of us no longer live in natural settings, and even those who do, modify them with such artifacts as fire, weapons, and tools.

Today, all of mankind, some to a smaller but most to a larger extent, make our own habitats, and our man-made environment becomes more and more complex. Moreover, in the process of adjusting to our increasingly artificial surroundings, we become less and less fitted to live in natural ones, and so the course upon which we are embarked is far more likely to continue than to be reversed. For Homo neocorticus we can project a habitat of, to us, incredible complexity, and we can imagine that this will provide the spur that could eventually produce the Group-brained neocorticus, whom we could no longer designate as Homo.

At this point we pass our project over to our reader to use his or her imagination to visualize a meeting between Homo sapiens and Group-brained neocorticus, or its prototype as it may exist somewhere in the vastness of the universe.

Sources

Amoure, John E., Johnson, James W., and Rubin, Martin. "The Stereochemical Theory of Odor," *Scientific American,* 210:42–49 (February 1966).

Beebe, William. *Half Mile Down.* New York: Duell, Sloan and Pearce, 1951.

Bennett, M. B. L. "Electric Organs." In W. S. Hoar and D. J. Randall (eds.), *Fish Psychology,* Vol. 5. New York: Academic Press, 1951.

Bovet, J. "Ein Versuch wilde Mäuse auf Hemmelsrichtungen zu dressieren," *Zeitschrift für Tierpsychologie,* 22:839–859 (1965).

Bullock. T. H. "Species Differences in Effect of Electro-Receptor Input in Electric Fish," *Brain, Behavior & Evolution,* 2:85–118 (1969).

Capart, Jean. *Thebes.* London: Allen & Unwin, 1926.

Clarke, Arthur C. *Profiles of the Future.* London: Gollancz, 1962.

Cooper, Max D., and Lawton, Alexander R., III. "The Development of the Immune System," *Scientific American,* 231 (November 1974).

Cousteau, Jacques Yves, and Diolé, Philippe. *Octopus and Squid: The Soft Intelligence.* Garden City: Doubleday, 1973.

de Vries, A. L., and Wohlschlag, D. E. "Weddell Seals," *Science,* 145:292 (1964).

Dijkgraaf, S. "Honeybees," *Zeitschrift für vergleichende Physiologie,* 30:252 (1943).

Dolhinow, Phyllis (ed.). *Primate Patterns.* New York: Holt, Rinehart & Winston, 1972.

Dröscher, Vitus. *The Magic of the Senses.* New York: Dutton, 1969.

Durant, Will. *The Story of Civilization,* Vol. I. New York: Simon & Schuster, 1935.

Erman, Adolf. *Life in Ancient Egypt.* Trans. H. M. Tizard. London: Macmillan, 1894.

Estes, R. D. "The Role of the Vomeronasal Organ in Mammalian Reproduction," *Mammalia,* 36, 3:315–341 (1972).

Frisch, Karl von. *Aus dem Leben der Bienen.* Trans. Dora Ilse. London: Methuen, 1954.

Frith, H. J. "Incubator Birds," *Scientific American,* 201:52–58 (August 1959).

Geldard, Frank A. *The Human Senses.* New York: Wiley, 1953.

Griffin, Donald R. "Echo-Ortung der Fledermäuse," *Naturwissen schaftliche Rundschau,* 15 (1962): 169–173.

Gruner, Hans-Eckhard. *Leuchtende Tiere.* Wittenberg: Neue Brehm Bücherei, 1964.

Heran, Herbert. "Untersuchungen über den Temperatursinn der Honigbienen," *Zeitschrift für vergleichende Physiologie,* 34:179–206 (1952).

Hertel, Heinrich. "Struktur, Form, Bewegung," in *Biologie und Technik.* Mainz: Krausskopf Verlag, 1963.

Hoffer, A., and Osmond, H. "Olfactory Changes in Schizophrenia," *American Journal of Psychiatry,* 119:72 (1962).

Hogue, Charles L. *The Armies of the Ant.* New York: World, 1972.

Hopkins, Carl. "Sternopygus," reported by Jane E. Brady, *The New York Times,* July 6, 1972.

Kaiser, H. E. *Das Abnorme in der Evolution.* Leiden: Brill, 1970.

Kellogg, W. N. *Porpoises and Sonar.* Chicago: University of Chicago Press, 1961.

Kolb, Anton. "Wie orientieren sich Fledermäuse während des Fressens?" *Umschau,* 65:334–335 (1965).

Jonas, David, and Jonas, Doris (Klein). *Man-Child: A Study of the Infantilization of Man.* New York: McGraw-Hill, 1970.

Jonas, Doris, and Jonas, David. *Young Till We Die.* New York: Coward, McCann & Geoghegan, 1973.

Lilly, John C. *Man and Dolphin*. London: Gollancz, 1962.

———. *The Mind of the Dolphin*. Garden City: Doubleday, 1967.

Lissman, H. W. "On the Function and Evolution of Electric Organs in Fish," *Journal of Experimental Biology*, 35:156–191 (1958).

Lorenz, Konrad. *Studies in Animal and Human Behavior*. Vol. 1. Methuen, London, 1970.

Lüscher, Martin. "Air Conditioned Termite Nests," *Scientific American*, 205:138–145 (July 1961).

Mangold-Wirz, Katharina. "Quelques problèmes actuels de la teuthologie méditerranée," *Rapport et Procès-Verbaux des Réunions de la Commission Internationale pour l'Exploration Scientifique de la Mer Méditerranée*, 14:1959. Paris.

Marshall, N. B. *Tiefseebiologie*. Jena: Fischer, 1957.

Merkel, F. W., and Wiltschko, W. "Magnetismus und Richtungsfinden in zugunruhigen Rotkehlchen," *Die Vogelwarte*, 23:71–77 (1956).

Milne, Lorus, and Bolle, Fritz. *Knaurs Tierreich*. Munich: Drömersche Verlaganstalt, 1960.

Moehrer, F. P. "Bildhören—ein neuentdeckte Sinnesleistung der Tiere," *Umschau*, 60:673–678 (1960).

Moller, P. "Communication in Weakly Electric Fish *(Gnathonemus niger)*," *Animal Behavior*, 18:768–786 (1970).

Packard, Andrew. "The Behavior of *Octopus vulgaris*," *Bulletin de l'Institut Océanographique de Monaco*, Special 1D (1963).

Roeder, K. D. "Moths and Ultrasound," *Scientific American*, 212:135–148 (April 1965).

Romanes, George J. *Mental Evolution in Animals*. London: Kegan Paul, 1885.

Scheich, Henning. "Neuronal Analysis of Wave Form in the Time Domain: Midbrain Units in Electric Fish during Social Behavior," *Science*, 185:365 (1974).

Seton, Ernest E. Thompson. *Life Histories of Northern Animals*. London: Constable, 1909.

Solecki, Ralph S. *Shanider*. New York: Knopf, 1971.

Sources

Wells, M. J. *Brains and Behavior in Cephalopods*. London: Heine-
     mann, 1962.
Wilson, E. O. *The Insect Societies*. Cambridge, Mass.: Belknap Press,
     Harvard, 1971.
Woolley, Sir Charles Leonard. *The Sumerians*. Oxford: Oxford Uni-
     versity Press, 1928.

Index